ETERNAL WOOL

ETERNAL WOOL

by Evelyn Alexandra Domjan

Domjan Studio

Peggy Brogan Editor

Library of Congress Catalog Card No.: 80-65727

©1980 by Joseph Domjan

ISBN 0-933652-15-1

Printed in the United States of America
Domjan Studio, Tuxedo Park, New York 10987

"Kingdoms have fallen, dynasties have disappeared through centuries of change. The great patrons who ordered them are gone long ago — but the tapestries they left behind glow in eternal splendor."

List of large-one-man shows of Domjan tapestries:
Laguna Gloria Art Museum, Austin, Texas
U.S. Capitol Bldg., Washington, D.C.
New Jersey State Museum, Trenton
The Cincinnati Art Museum, Ohio.

Joseph Domjan, the world renowned "Master of the Color Woodcut" is a prolific artist far-famed for his achievements in the color woodcut. He developed a unique style so rich with color that his works appear as oil paintings.

Domjan was born in Budapest in 1907. Orphaned at an early age he learned a trade and went to work to help support the family. During the depression of the 1930's unemployment sent him to other fields and he became a trained weaver. His natural artistic talent gradually emerged and he designed fine upholstery fabrics and wall coverings until economic hardship terminated his business. He then spent several years walking through most of Europe, selling sketches for a meager living, and studying great works of art in museums. He ultimately decided to have a formal art education and graduated with highest honors from the Academy of Fine Arts in Budapest seven years later. It was at the Academy where he met his future wife, Evelyn. After the end of World War II he launched his career in Scandinavia, followed by successful exhibitions throughout Europe, Russia, and as far away as China and Mongolia.

Domjan, believer in the concept of "Life, Liberty and the Pursuit of Happiness" realized the ambition of his adult life on August 31, 1957, when he first set foot on American soil. He had his first one-man show in New York the same year. Numerous other exhibitions followed throughout the United States of America of color woodcut, oil paintings, and starting 1968, of tapestries. To date the artist has three hundred fifty one-man shows to his credit and one hundred fifty museums own his work. His tapestries represent the culmination of many years' artistic maturation.

This book is dedicated to the splendor of Domjan's tapestries.

CONTENTS

Tapestries of Long Ago

Lamb   *Domjan*

Tapestries radiate pomp and splendor in royal ballrooms, grand opera houses. As a means of commemorating historical events, persons, noble tapestries command every eye and imprint lasting memories. Tapestries, old and new, are the ultimate wall decor. Movable frescoes, tapestries are the bridge between wild nature and a man-made world; in modern buildings, tapestries become part of the architecture, softening hard surfaces of concrete, brick, marble, and create a wall of enchantment, beauty, harmony.

In tapestry-making alone there is an unbroken tradition, a continuation of techniques, of weaving methods practiced since the dawn of history. Almost in the same way the shuttle passes under and over a set of parallel threads today as in the hands of the slaves of Nineveh, of the Coptic monks, of the medieval artisans; in Arras, in Tournai, at the royal workshops of Les Gobelins in Aubusson and in workshops right down to our own times. This continuity of method and technique is all the more precious today since so many old techniques in other fields have died out. Despite our scientific advances, we have been unable to discover the formula and produce stained glass windows like those of Chartres Cathedral, or the cloissone enamelings of early Limoges. But look at the works of modern tapestry-makers. The modern French tapestries interpret the artist's vision with technical perfection. A specialization of trades in olden times resulted in families pursuing the same trade for generations; weaver father and son, to a tradition shared experiences carried forward, continually refining, perfecting. Aubusson for example, has its families of weavers today.

The basic material for tapestries is wool; wool is soft and strong at the same time, permissable and lasting; a warm, living material. Its dyes are permanent. Tapestry in its present processing can withstand heat and dry air, humidity, draft, storage or exposure to light and remain unchanged for centuries to come.

Tapestry today commands high prices, though the materials used are still relatively inexpensive. The value of the tapestry lies in the beauty of the design, the force of creation of the sketch or more finished art work which then is made into a cartoon; and the perfect execution, the virtuosity of the weavers. There is a transfer from one technique to the other — only in extremely rare cases does a tapestry start at the loom without a sketch or cartoon. And there is a collaboration between the designing artist and the artisan weaver or weavers who realize the work.

In its subject matter tapestry can be historical, religious, genre-picture, land-scape, still life — even portraits were woven during its history — or tapestry can be abstract, ornamental, symbolical — that is, tapestry can be everything; but naturalistic portraits, "tromp d'oeuil" effects are not fitting to the character of the tapestry. Only when tapestry attempted to be painting did such mistakes occur.

Great ages of tapestry coincide with great ages of architecture — Gothic Baroque and contemporary Modern.

In the early glory of tapestry making, when workshops in Arras and Paris, produced immense quantities of the most magnificent tapestries (many of which are preserved in museums, with colors still brilliant despite adversities and centuries of neglect) tapestry was *the* major art form excelling over illlumination, painting, sculpture. The 17th century which raised painting above every other art form considered tapestry merely as an imitator of painting. Thus tapestry came to be regarded as an applied art. During the last thirty years, however, the position of tableaux painting, unchallenged for three hundred years, has been destroyed by painters themselves in search of new art forms. Revitalized, modern tapestry has once more reoccupied its position of leadership. This movement has spread from France, where the taste for tapestries was never lost.

When you are alone, tapestries are good to meditate or dream by — they become your companions. When entertaining friends, they create an atmosphere of cordiality. Tapestries are a suitable background for singing, dancing, celebrating, for they are festive. It is good to listen to music with tapestries surrounding you. Not only are they a good decor, but tapestries make for excellent acoustics. Tapestries are like a magic carpet to carry you away into the land of dreams.

Le Corbusier called tapestry the nomad's mural. He was not thinking of Bedouin tribes and their tents, but of modern men in cars, airplanes, moving from one city to an other and frequently changing one impersonal apartment or hotel room for another. He envisioned prefabricated apartment units with most of the furnishings built in, tenants arriving with only their suitcase, some books, records and of course their own tapestries. Tapestries easily transported, (care-free) unfolded, hung on the wall mysteriously change an impersonal room into a very unique, personal environment.

Lurcat called tapestry the skin of the wall: warm, pliable and good to touch, alive. In Aubusson, when the weavers speak of the intensity of color, the brilliance of a tapestry, they say it has "warmth," and "fire" like the blood, like the heart.

Walls enclose, deprive us of sunshine, green foliage, bird song, starlit nights, the sea, the sand. A tapestry bring these things back to us, the color, the music, the movement, the rhythm, the texture. The woven thread always casts a little shadow—weaving i s three-dimensional. The repetition of parallel lines of color transitions is like the beach when the tide has gone out. The color doesn't rest on the surface as with paintings, but penetrates, is there in thickness even in reverse. And, surely enough, who can resist folding back a corner to look at the maze of wool on the reverse side of the tapestry, only to enjoy the "right" side, its image and color. all the more.

Ptah, God of Artisans   Domjan

Prehistoric Origins. The practice of weaving goes back into the remote past and it is difficult to tell whether fingers developed more and more advanced techniques of weaving or if the exercise of weaving made the fingers more and more skilled.

During the Neolithic period, about twelve thousand years ago, with the collecting and storage of food, containers were needed. The first containers were baskets woven of branches, reed and grasses, made by the women who stayed near the hut with the children.

Magical ceremonies called for costumes; grasses, shreds of bark, rushes were at first woven into fabric and used as ceremonial dress even before yarn was spun. Only later were woven fabrics used as cloth to cover the person as a protection against the weather. Fashion throughout history served first as adornment, its practical uses often forgoten. Since earliest times woven fabrics were dyed and decorated with shells, feathers.

Paleolithic culture centered around fishing and the gathering of shells. Twisted and spun fibers were made into cords, having had many uses. Netting, knotted and looped nets are also forerunners of weaving techniques. The picture of fishermen mending their nets, nets drying on the shore, across their lacy pattern the immense blue sea, is a picture of unbroken continuity reaching beyond the dawn of history.

According to an old Potawatami (Wisconsin) Indian legend, an old woman in the moon is making baskets; when she is finished the world will come to an end; but a dog always eats her baskets (the moon eclipse) and so she always has to start again.

Baskets are the earliest human industry made by hand without a frame or a loom. While flint and pottery fragments turn up frequently, baskets and cloth disintegrate in the soil; in very exceptional cases an imprint may remain. Many different plants were used for baskets; branches of the willow tree, reed, palm, bamboo and roots. Shelters, huts, mats, canoes were made of baskets. A great variety of early objects from cradles to coffins were baskets. The method of basketry found in Etruscan tombs is the same technique used by primitive tribes today.

There is a connection between baskets and pottery—which came first? American Indians lined their baskets with clay; on an Andaman island, on the other hand, a basket protection was woven around clay pots. Moses was put into a basket coated with mud.

Baskets, netting—do they belong in a history of tapestries? Not if we think of the development of classic tapestry and Les Gobelins, but there has been much experimentation recently with knots, netting, combined with weaving. The result: transitional art works, textile hangings and soft, mobile sculptures. This newly discovered art with yarn uses every known and forgotten technique of weaving.

The First Looms and Fibers. Weaving was first made without a loom. Parallel reeds, grasses or yarn was placed on the ground—the warps. At right angles another, the weft, was passed under and over warps alternately and on the return under where it went over and vice versa. This is the basic system of weaving and mending, a process invented by many different peoples and carried

Nephertiti and the Weavers   Domjan

over to our days from prehistoric times. Weaving is fundamental and preceded civilization. Cloth served for a great variety of purposes including bags or sacks for carrying things. Real weaving started when fibers were spun into yarn. The yarn can be made both of plant and animal fibers.

The Sumerians replaced the sheepskin they had been wearing by woven cloth and used wool fibers without spinning. Vegetable fibers are the flax, cotton, hemp; animal fibers are the hair of the sheep, goat, camel, and the long-haired lama, vicuna and alpaca of Peru. Fibers suitable for weaving are strong, elastic and will take dye readily.

The ancestor of the loom was a branch of a tree, a horizontal branch to which the yarn was fixed. The warps were hanging down vertically and the weft passed under one and over the next horizontally. The next step was to attach stones or bricks to warp threads.

Neolithic lake dwellers used a perennial flax. An annual flax was grown in Palestine five thousand years B.C. and bleached to white in the sun.

Cotton comes from India. Pliny writes this charming tale about cotton lambs of Tartary: "These tiny little lambs are born in the field, they eat the grass around them and when the grass is gone, they look for other pastures, but leave their wool-hair behind." The Indian cotton has short fibers, it is soft and light and its color is cream to white. Banana leaf fibers and pineapple fibers were also used.

The most precious of all fibers is silk and there is no end to the legends connected with it. Silk fibers derive from the cocoons of a moth. Wild cocoons give short fibers and only the delicate domesticated cocoons first produced in China can be reeled.

Tapestry weaving uses wool for the weft threads and cotton or hemp for the warps. Silk and metallic gold and silver threads were widely used for the weft from the Renaissance period onward. Modern tapestries are woven with wool.

A more advanced kind of loom was a wooden frame with warp threads hanging down. Etruscan and Greek looms were made this way. Later the parallel warp threads were fixed and cylinder rolls served to roll up the finished work and unroll more yarn.

Around 3,000 B.C., the draw loom, the plow, the sailboat, and the potters's wheel were all developed at about the same time as the calendar, astronomy, and an early form of picture writing; art of the highest order was produced during this early era of ancient civilization.

For a colorful fabric, cloth can be printed after the plain weaving is finished. Color can be applied on cloth from carved woodblocks. The surface of the block is inked, and by pressing the block face down on the cloth the pattern is transferred; the result is a printed textile. This method was used in India since ancient times. Woodblocks were also used for textile printing in Europe during the Middle Ages and in China. Another method of obtaining a colored fabric is by submerging the woven cloth in a dye. Different patterns and shades can be obtained by blocking out parts of the fabric that will not take the dye; the recently rediscovered "tie dye" method is actually an ancient Oriental method.

A different method is to dye the yarn in different colors before weaving, then

Etruscan Women Dying Wool Kroninger

weave it into colorful fabric. Tapestries are made in this way. The wool is dyed first and woven according to a pattern, called the cartoon.

Egypt. The dry, warm climate of Egypt has preserved wall paintings and bas-reliefs with their colors looking as fresh as though they were painted yesterday. The paintings show the gods, pharaohs, their ladies, priests, nobles, richly dressed. According to Pliny, Egyptian textiles surpassed the Asiatic in richness. Along the Nile, dyers knew how to produce nuances more delicate than flowers.

There were many different kinds of textiles in use; finest linen, seen blown in the wind, folded in small narrow pleats and worn by princessses; stiff heavier ceremonial garments of the priests; curtains; ornate sails of pleasure boats on the Nile. From swaddling cloth to mummy wrappings, textiles were used for many things. Weaving was a highly developed craft and art. Nails were not yet invented; cords were used instead to make furniture. A magnificent throne chair of the Pharaoh Tutankhamen is decorated with gold plaques. The legs of the chair are fixed with cords.

One of the oldest varieties of fabric is linen, threads of which are prepared from the fibrous parts of the flax plant stalk. Methods and processes used in the cultivation of flax; spinning the fibers into thread, the manufacturing of linen, weaving are well illustrated in tomb paintings. Slender slave girls are beating the flax, combing it, spinning the yarn and weaving. One weaver works at an upright (high warp) loom. The warp threads are fixed to a fan-shaped frame. Some fabrics have woven or embroidered motifs; symbolic animals, plants in yellow, green, red, blue and brown colors. Fragments of the lotus garment found in the tombs of Thutmose III (1400 B.C.) at Thebes are richly decorated textiles. They have repeated smaller and larger lotus plants in a symmetrical checkerboard pattern, a border of papyrus plant motifs and fragments of the cartouches of Thutmose III and Amenotep II of the 18th dynasty (1555-1350 B.C.).

In the Metropolitan Museum of Art there is a model of a weaver's workshop from Girga, 12th dynasty. The stuccoed and painted wood model shows a large simple hand-operated horizontal, low-warp loom. A female slave is winding the yarn onto a bobbin. A male slave is the weaver, two others, crouching at both sides of the loom, operate the batten, a wooden bar that changes the position of the warp threads. The shuttle is passed between by the weaver.

Through several thousand years of ancient Egypt, and during the times of the Roman occupation and early Christianity, looms were busy in Egypt; then came the Coptic tapestries.

Weaving In Antiquity. Weaving was highly developed among the Etruscans, who originally lived between the Tiber and the Arno during the 9th - 8th centuries B.C. Theirs was an urban civilization. During the 7th - 6th centuries B.C. a specialization of industries took place; their artists-craftsmen produced magnificent bronzes, terra cotta figures, small statuary; they had their own writing. Etruscan autonomous city-states built on mountain tops extended their political and commercial control over almost all of Italy. They waged sea battles against the Greeks. Trade routes over sea and land carried their goods North and East until Rome gained power. Among these goods, we assume, were textile products.

19

Etruscan Women Weaving

Etruscan women produced richly colored woven and embroidered textiles as seen on wall paintings and scenes painted on amphoras. Surviving works show two women standing at an upright loom, one pushing the shuttle, the other directing the threads; two other women are shown folding large pieces of textile while another is seen holding a scale. These women are at work, while others are dancing or performing religious ceremonies; all are dressed in long gowns of graceful folds decorated with geometric patterns, stars, palm motifs, the meander.

*   *   *

Walls of the palace of Nineveh, destroyed in 606 B.C., were covered with alabaster slabs in colored bas-relief. Figures in long tight robes that are patterned and fringed, engage in religious ceremonies. Their sumptuous garments are decorated with woven motifs of winged animals, bulls, lions — the sacred tree, the pomegranate. Ancient texts often use the word "broidered" in descriptions, but the word has uncertain meanings. The ornamentation of textiles may have been embroidery, or a compromise between darning and weaving supplemented with needlework.

Pliny writes about the rich carpets of Babylon, perhaps tapestries. Sassanian kings (250-650 A.D.), like the Persians before them, lived in great luxury. Weaving was on a high level. From Mesopotamia weavers came to the royal Sassanian workshops in Susa; Byzantine weavers also came to work for the Sassanian kings, and so old Sassanian and earlier Persian motifs were enriched by Greco-Roman motifs. A typical Sassanian subject was the hunting prince in a medallion, small circles with wild animals around him. Pairs of beasts; griffins, lions, eagles, deer, ostriches were also favorites, pictured facing each other under the sacred tree, life-tree.

Through the history of tapestry, weavers migrated to other countries because of war, for better pay or other reasons. Motifs migrated with them. One particular motif is rarely limited to one area. Sassanian lions and eagles make their way into heraldry and heraldic tapestries. Tapestries, most mobile of art forms, or fragments of textiles were carried away, incorporated in altar cloth, church robes.

Greece. During an excavation under the church at Lacco, on the island of Ischia, weights of a Greek loom were found and the upright loom reconstructed. The vertical, high warp loom had no frame on the bottom to which the warp threads could be fixed; they were straightened by weights. The wefts were pushed upward and the work proceeded from the top downward. The Greek shuttle was a short rod. Since the weights in Ischia were found next to funeral urns, the loom had been buried with its owner, a woman, from the times when the Greeks controlled the sea routes and colonized the southern shores and islands of the Italian peninsula, around 500 B.C.

Classical Greece was a masculine society. Woman's place was in the home. There, in protected seclusion, the lady of the house supervised slaves who prepared the wool, spun, wove and made cloth; thus providing the household with the needed textiles.

Tapestry weaving was certainly practiced in ancient Greece. Not many textiles survive, but fragments from Greek tombs were found from the fifth to the first century B.C. that came to light as far from Greece as the Crimean

Penelope

peninsula. We can discover more from literature. Realistic leaf, laurel, ivy, acanthus scrolls were woven; animals, ducks, birds, dolphins show Asiatic influence. The colors were vividly described; violet blue, purple, green, saffron.

Greek mythology has its stories of weaving; Arachne, Neonia and Minerva are weaving in competition. In the Iliad and Odyssey, Trojan women are praised for their skill in weaving and embroidery. Andromache is working on embroidery when the news of Hector's death reaches her. Penelope spends her time at the loom while waiting for Odysseus's return. Her vertical, high warp loom is seen on a Greek vase of the fifth century, B.C. found at Chiusi. Here too, the warp threads are kept in place by weights. On the upper part of the loom is the finished tapestry weaving; a ribbon of winged horses and figures.

Rome and Pompeii. Weaving was first the occupation of women in the home. The Roman matron was sitting at the loom, her trained fingers handling the shuttle — a hollow bone — practicing weaving that, even then was already an old tradition, almost a way of life. The Roman *gynaecea* like the Greek *gunaikonites* were quarters set apart in the home for spinning, weaving, tapestry making. Tapestry wall hangings and couch covers with woven figures of mythology came both from private *gynaecea* and *collegia opificum,* workmen societies. Already in Greece, from the third century B.C., the weavers became professionals and formed companies. Brotherhoods, fraternal colleges existed among urban craftsmen of the Roman empire.

Wool making was a chief industry of Pompeii. Wool scourers, weavers, fullers, dyers, felters, were differentiated craftsmen with large workshops at different places in the town. The owners of fulling mills had an important position in the economical and political life of Pompeii. The "fullones" was a powerful association.

At the time of the last days of Pompeii, before the town was destroyed by the eruption of Vesuvius in 79 A.D., the fine private homes built for graceful luxury show signs of decline. The houses were built with atrium, peristylium, walls painted in red, yellow, black or ivory. Red columns enclosed enchanting gardens with vines, trellises, flowers and fruit trees; statues and fountains.

The walls were painted with an easy routine. The painted decorations first accentuated the structural elements of the building but, as the unity of the building was lost, painting became louder in colors and no longer followed the architecture. Complex, imaginary construction were painted; pillars, columns, stairs, terraces in perspective: curtains with tomp d'oeuil, behind them fanciful gardens, sea-scenes carry the viewer far beyond reality. The painters followed the wishes of the owner-patron and the theme of the frescoes embraces all activities of life: religious scenes, funerals, banquets, the participants painted as individuals, portraits, gestures well observed. Bacchus is depicted with grapes, behind him the beloved Vesuvius. On the frescoe of an inn men are playing with dice.

The villas of Pompeii were pressed by urban development in the crowded, walled-in city. Only few remain in their original elegance while most belonging to rich merchants became an agglomeration of shops, workshops, living quarters for family and slaves. The traditional dignity of the patrician home was sacrificed to trade and commerce.

Marcus Vesonius Primus was the owner of a fulling mill. His workshop was installed in a converted private home. He was active in politics, campaigning for votes for his favorite candidate. He put his name and those of his workers as supporters of the said candidate. His fulling mill occupied the once beautiful peristyle with three basins of running water. The wool was washed in washing pans containing fullers' earth (long before soap was invented). Murals show realistic scenes of the working processes; a man is standing in a stone vat, washing, scouring the wool. Another man is shown cleaning the wool with brushes as it hangs from a frame. The bleaching is being done by exposing the wool to sulphur fumes. Marcus's neighbors surely had good reason to complain about air pollution in Pompei one thousand and nine hundred years ago.

Pressing was done in a press similar in principle to the one used today by bookbinders. The products were both produced and sold under one roof. Next to the vestibule was the store to the street with a portrait of the proprietor, a painted shop sign. Upstairs were the living quarters for the family.

The largest dye workshop was installed in a peristyle along the Via Stabia and had nine brick furnaces furnished with leaden pipes to boil the water. Fabric for the expensive purple togas worn in Rome was produced here.

The textile plants of Vecilius Verecundus, like others were also a lucrative business in this bustling trading town. Here there were three open shops and the painted shop signs show women standing in the shop and showing textile products to the customers who touch the fabrics and bargain. The open shop is shaded by a wide over-hanging roof, bedrooms upstairs had more room this way — it was a frequent solution in the crowded city. Above the shop signs religious scenes are painted; Venus, patron goddess of Pompei, is shown in a quadriga or wagon pulled by four elephants. The god of commerce, Mercury, comes out of a temple-cell with a purse full of money "Salva Lucrum."

All weaving was done by hand, but even so, there was already a separation of a growing textile industry and the country home looms where there was more leisure time for handicraft, and a patterned weaving — made with patience, and love for their own use and not for sale. A talented woman of the household, or a slave, perhaps from the Orient, working at curtains for the household — it is here where we find the origins of tapestry weaving.

We have left the Via dell'Abbondanza and the shops and shop signs, passing by the large oblong building of the cloth-exchange, erected by a priestess named Eumachia. This building has an open court, surrounded by a colonnaded walkway on four sides. The sun goes down, its last rays illuminate the temple ruins and the triumphal arch of the Forum — across the arch rises Vesuvius in golden glory against the violet sky.

Coptic tapestry was produced by the Copts who were the native Christians in Egypt after the Ptolomais period and during the rule of imperial Rome; it continued, but declined slowly after the Arab Islamic conquest of 640 A.D. The word "Copt" derived from the Greek Apigytos — this became Kipt in Arabic and Copt in Latin. The Copts were of Hamitic origin, like the ancient Egyptians: they were not Greeks, nor Romans, nor Arabs. Coptic art produced small size tapestry pieces. There for the first time in history is a rich, fully developed art we can see and enjoy — not only guess after literary sources — images are woven on a loom; portraits, scenes of figures, animals, ornaments, and there are a lot of them. Had we not seen them behind glass among the treasures of museums from the Kremlin to the Victoria and Albert Museum, but in an art show today, these small bright tapestry masterpieces could deceive us into thinking they were avant garde experimental work. Giacometti once said that if he were to make tapestries, he would make them very small, 5 x 5 cm. I could rather imagine a Giacometti tapestry 5 cm. wide and 8 feet high. Other artists had another opinion of tapestries. Modern tapestry is measured in feet rather than inches and is meant to hang on a wall.

Coptic tapestries are garment trimmings made to decorate tunics, sewn on or appliqued on plain linen-woven garments of priests, and for funerary-burial garments. There are shoulder-stripes, 1-6 inches wide; neck-pieces, and small square panels and round medallions, 30 x 30 cm.; here are also little square pillows. All are made with coarse woolen wefts on linen warps, similar to the technique used today in Aubusson and elsewhere. Tapestries have 12 warps and 27 wefts per cm.; 9 warps and 24 wefts per cm.; and in the last period 6 warps and 14 wefts per cm.; the technique became coarser instead of getting more refined by evolution. A characteristic of earlier coptic tapestries is the oblique weave; the wefts are slanted in the direction of the design instead of going at right angles to the warp threads.

We look at an early piece of Coptic tapestry; this is a woven portrait of a woman who looks back at us with large dark and sad eyes. The face, the hair, is shaded in the naturalistic late-Hellenistic style. The tapestry has a narrow border of pearls — Oriental influence. Another tapestry depicts the God of the Nile; gods and goddesses of Antiquity are portrayed in the Roman manner; a fine hatching is used.

Later pieces are more decorative, a strong simplification results in a purely ornamental style, hatching is no longer used. Figures are out of proportion, distorted with large head; patches of plain color give an abstract pattern. During the last period there are scenes woven with many small figures; they remind us of tapestries made by Peruvian Indians in a later period.

Strong monastic communities were isolated sanctuaries of culture in Egypt. Icons, illuminated manuscripts were painted by the devoted monks who also wove tapestries. This culture was flowering at the time of the Arab invasion.

Byzantine, Persian, Syrian influences, Sassanian motifs mingled with Christian symbolism and Egyptian traditions. It was an independent art, ornamental; naive and sophisticated combining perfection and ignorance in the shadow of a great past.

Christian art, especially at its earliest period, used symbols derived from the bible or apocryphal scripts, rather than the human figure. A favorite symbol was a lamb on the mountain or paradise; the fish, the cross, and monograms of Christ were used; also vine, grapes, and the amphora. The lion, the griffin, and the phoenix indicated the coming of the Messiah. The pomegranate, ancient fertility symbol came to symbolize the suffering of Christ. The hare was a reminder of the resurrection, here memories of the Egyptian Osiris cult and the Roman Dionysus mysteries echoed. ☥ the cross of the Pharaoh's, the so called "key of the Nile" was considered to be a pagan fore-telling of the coming of Christ.

On one tapestry a hare is eating grapes. The small square has a border of ornaments. A shepherd is milking a goat under a trellis of vine; the border is of facing pairs of doves. Another square tapestry has a center medallion with an apple tree — symbol of paradise. The background is brown, strewn with small figures riding on phoenix birds, with fish swimming between them. Here and there is the motif of the cross. Figures, fish, phoenixes, have a black contour line. The colors used are red, blue, and the purple obtained by a dye of madder and indigo. Colors are clear local colors, used in a decorative way, the fish is red, the figure blue, the spacing of motifs textile-like. There is no attempt made to imitate reality. One may think for a minute of Miro or Kandinsky. Seen in reproduction the small pieces seem large. They are monumental. For their unique style, simplicity, decorative qualities, coptic tapestry is unsurpassed.

Medieval Tapestry Workshops

Nuremberg Dürer

Spring on the Way Domjen

Technique of Tapestry Weaving. Everybody likes to lift a corner and peek behind a tapestry. The first question asked is: how was it made? Tapestry is woven on a loom. The basic technique of tapestry-weaving has not changed since antiquity. In Babylon, Assyria, Persia, China, and in Byzantinum, walls were decorated with textiles. In Ravenna, on the mosaic freeze of San Appolinare Nuovo, we can see in the palace of Theoderick of the sixth century tapestries hung between columns to enclose rooms. Different cultures, far from each other in space and time each discovered weaving independently of the other. Among the Incas in Peru the dry climate preserved mummy wrappings in perfect bright colors; animals, heads of warriors, meanders, zig-zag lines, birds, geometric ornaments were woven, and the textiles were buried with the dead with jewelry, fruit, and weapons.

The so-called Bayeux Tapestry, the Norman story of the conquest of England from the eleventh century is not a tapestry at all but an embroidery. It is needle work on linen.

In France, already King St. Louis encouraged the artisans. In Arras, Mahaut, Countess of Artois, was the great art patron. The dukes of Burgundy, the French court, with their patronage, and love of art, produced the golden age of tapestry.

Ever since then tapestry has been a part of life in France; so much so that it is included in popular expressions. "Faire la tapisserie" literally means to make tapestry. Idiomatically it means to be a wallflower or unpopular young girl at a dancing party. "Regarder par les trous de la tapisserie" literally means to look through the slits of the tapestry or idiomatically be an outsider.

Tapestry weaving is done on two kinds of looms in use since the Middle Ages. High warp or vertical looms (haute lisse) and low warp or horizontal looms (bass lisse).

Between two poles parallel strands of threads are stretched; these are the warp threads. The weaving work is to cover the warp threads by another set of threads, the weft threads completely, by passing them one by one alternately over and under the fixed threads and in reverse on the return, after the end of the line, the selvage has been reached. In case of a simple linen weave without a pattern the weaving goes on uninterrupted from selvage to selvage. The tapestry weaver, on the other hand, works on small color patches. His weft threads are of colored wool. He is following the design. Assuming that he is weaving a blue garment of one of his figures in the tapestry, darker and lighter shades of blue are marked on his cartoon, or painted; these are the drapery folds that follow the shape of the figure. The weaver works with a dark blue for a while, then lets the bobbin that holds the dark blue wool hanging and works on a lighter shade; he will pick up the darker shade again when he gets to another patch of dark; he may dovetail these and change colors line after line on his small patch for fine modeling. Little by little, row after row, the tapestry is growing.

The artist, when designing a tapestry, must keep in mind that the warp threads are at right angle to the weft threads. Also, when a contour line makes a diagonal direction the weaver can follow this only in a step by step zig-zag line. The tapestry should have bright colors and clear contour lines. The limitations of the techniques were refined to such an extent that tapestry lost its character. Modern French tapestry — the production of the last twenty-five to thirty years has resulted in a new blossoming of an ancient art. So we can talk about tapestry that is at the same time very old, a heritage of glamorous centuries and a very young and vital art still bringing surprises, new experiments and great possibilities to the contemporary art scene.

Thick threads and a wide distance between warp threads result in a coarser type of tapestry. More warp threads per inch and thinner weft threads result in a finer weave which allows for more detailed work. The artist and the weaver have to find the right balance. Coptic tapestry had only six warp threads per cm. in its last period; the Apocalypse at Angers had only five. During the eighteenth century there were ten and twelve warp threads per cm. Today four to six are used per cm. by most artists.

Fine craftsmanship can be recognized by even, straight and tight rows of weaving. An important technical question is the graduation; the blending of one color into the other. In this there have been different methods at different times and according to workshops. The simplest method is to weave a few rows in one color, then next to it, a few rows of a darker or lighter shade of the same color. Earliest European tapestries, like the Halberstadt tapestries were made this way; at that time, the 10th and 11th century, the weavers did not know better. A more advanced technique is the hatching when lines of one color are dovetailed into another one; this gives a fine modeling and is a basic method in tapestry. Middle Age tapestries used it, modern tapestry uses it, the Domjan tapestries use hatching. Certain workshops had their own method of hatching; long vertical lines were used during the 15th century; Bruxcelles tapestry have a special short hatching by which they can be identified. To imitate painting, hatching was abandoned during the 18th century in French workshops. Besides hatching "la method chiné" permits a subtlety of graduations and values, also used in the Domjan tapestries. "Le piqué" is a method of using wools spun of different shades of grey for the modern black and white tapestries, for instance, or pre-blended colors for a tapestry in color to bring a changed effect and eliminate time and work spent on hatching. It fits well on non-figurative tapestries. If the wool is of an uneven thickness this gives another dimension and texture to the tapestry.

The work starts with a small sketch, the "petit patron" or "maquette". This is enlarged to the size of the future tapestry by the cartoniers, makers of the cartoons. During the Middle Ages cartoniers had their own style; this was the right style for tapestries. Cartoon makers were paid several times the fee given for the sketch. In other instances, as probably in case of the Angers Apocalypse, the artist made the sketch and the cartoons. LeBrun made his cartoons, and before him Raphael, and so do many of the modern artists.

The cartoon is finished. Next comes the dyeing of the wool, according to the sketch. The colored yarn then is put on many bobbins. The Medieval tapestry used 12 - 24 colors; by the end of the 18th century there were tens of thousands of shades. Now we are back to a limited number of colors; the Domjan tapestries use 20 - 28 colors. Setting up the loom for the new tapestry is an important part of work. The tapestry can be as high as the loom is wide and much longer (wider) because most of the warp threads are rolled up; the design is always woven at right angles. A standing figure appears in a horizontal position on the loom and the horizontally worked row-after-row of hatching, once the tapestry is hung on the wall, is a system of vertical lines.

High warp, vertical looms are used in Les Gobelins today. Two upright poles are holding a cross-beamed warp-roller around which the warp threads are wound. Warp threads are fixed to a rod that is inserted into a groove in the roller. The rollers and the supporting uprights must be very strong to take the pull of the threads. In case of a large tapestry woven at Les Gobelins, for instance, each thread has to take the pull equivalent of three kilograms. Counting 60.000 meters of thread the total tension of the beams would be 16 tons. The loom is fixed to the floor and the ceiling. There can be up to eight weavers sitting at a large loom side by side, mostly there are two; a smaller tapestry may be made by one weaver alone. The even and odd sets of warp threads are separated by a wood staff — "baton de croisure" and operated by strings called "lisses" over the head of the weaver. The wood staff, 40 cm. long at both high warp and low warp looms, facilitates the passing of the weft threads between the tightly drawn warp threads.

The weaver holds the pin in his right hand; this round spindle, shuttle, or pin; holding the weft thread, the dyed wool, is of different shape, and is called different names in different workshops; "broche" in Les Gobelins, "flute" in Aubusson. The pin goes — this is a "demipasse", returns, "duite". At Les Gobelins the weaver pushes the rows together with the pin, the comb is used to correct a row. At Beauvais the ivory or steel comb, held in the weaver's left hand, is used all the time. The weaving is started at the bottom. Weaving upward, when the weaver reaches the limit of comfortable height, he releases more thread from the upper roller by opening a clutch, the completed part of the tapestry is lowered and rolled in on the bottom roller and the clutch is fixed again; the work continues.

In case of high warp looms the cartoon is placed behind the weaver on the wall. A small mirror reflects the part the weaver is working on. The maquette is there in a corner to check colors. The contour lines have been traced on the warp threads. The weaver is working on the reverse side of the tapestry, but he can go around the loom and see the right side of the tapestry, at least a part of his work.

It is almost impossible to tell if a tapestry was made on a high warp loom or low warp loom. Beauvais and Aubusson workshops use low warp looms. The low warp or horizontal loom is supported by four uprights, like a table. The rollers work as with the high warp looms. The difference is, that the shafts are worked by treadle, which, when depressed, cause the separation of even and odd threads, therefore the low warp technique is also named "tapisserie à pedales" or "tapisserie à marches". This technique leaves both hands of the weaver free and so the work is somewhat faster. The pedals were a great invention and the low warp looms are a later method. The low warp weavers have a detailed cartoon with all colors marked; this is placed flat under the warp threads. The weaver works on the reverse of the tapestry like the high warp weaver but in his case he can never see his work, only small parts of it in a mirror he holds from time to time between the warp threads above the cartoon. The weaver of the low warp technique is weaving a mirror image of the cartoon — this often resulted in heroes holding their sword in their left hand and texts written in reverse. Today the cartoons are made as a mirror

31

Early British Loom

image of the maquette.

In both cases, high warp looms or low warp looms, the great moment comes when the finished tapestry — result of many months of hard work — is taken down from the loom "tombeé de métier". At a few inches distance from the last woven row the warp threads are cut; the workshop gathers to see the new work, two weavers hold the tapestry in the right position; if the head of the workshop is satisfied, there will be a glass of wine.

There are still minor works to be done; In the tapestry there are slits between color patches. Nomadic tapestries have these slits left open for the light to come in between when hung as tent walls. Medieval tapestry, French workshops, modern tapestry has the slits sewn together carefully from the back by hand. Weft threads are also finished in the back. The tapestry is then stretched, ironed, and is ready to leave for an exhibition or to be hung at its final destination.

The tapestry can be hung on loops, rings, or there is a staff inserted in a ribbon at the back, by which it can be hung. Tapestry can be hung on a wall, or it can be a wall — in any case it has to hang freely and should never be framed. A slight undulation of the surface adds to its beauty.

Tapestry in War and Peace. In Medieval Europe tapestries were first used in churches and in monastic institutions for chancel screens to keep the draft away. Smaller pieces served for seat and (kneeling) cushions, dossiers, baldachins. The terms: insulation, sound-proofing, acoustics, were unknown in Medieval times, but such beneficial properties of the tapestries were well known and made use of. The medieval town house had a kitchen, shops, downstairs, with laundry and hot rooms, and upstairs the large main room—often over-hanging above the entrance—was divided into sections by tapestries. The areas were often named after the tapestries; foliage chamber, red room, lion room, etc. Tapestries covered tables, chests, even floors, in homes of rich burghers.

In the castles gray stone walls blosomed when hung with "milles fleurs tapestries. Inside large halls part of the interior was closed in with partitions of tapestries and a room created within a room. There was need for intimacy in the austere architecture. The tapestries created a unified, complete decor, even doors were masked—this was the total pictorial environment where one could enter and walk into the picture, be part of it and surrounded by it. It is inevitable that super skyscrapers must have inside rooms without windows; tapestries are the answer for such enclosures, large or small.

With medieval tapestry rooms there was a dangerous space between wall and tapestry; Hamlet kills Polonius behind the tapestry; Falstaff falls asleep behind the tapestry.

Tapestries had many uses; ladies of the court looked at tournaments under tapestry-decorated canopies. Tapestries covered the large traveling chests, were used for caparison for horses: Nicolas Bataille, Master Weaver of Paris, even had to weave the caparison for a favorite leopard.

The tapestries were easy to move; they were rolled up, folded, and taken with the mobile court. In a castle at arrival the tapestries were hung with large nails, and the scene was ready for a festivity, banquet. The princes traveled as much for pleasure as out of necessity; their power, authority was kept by personal supervision; tax collection, performance of justice, was so arranged. This lasted until the end of the 14th century, by which time archives, registers grew to such an extent, that affairs had to be arranged by agents of the government; relegated authority. Charles VI journeys can be followed by the accounts the record shows how much was spent on hooked nails "for hanging the King's Apartments" with tapestries. Queen Anne of Britanny had a train of mules; the tapestries were packed in coffers, loaded on pack animals, caparisoned in royal colors, and an official "driver of tapestry sumpter" was responsible for the safe arrival of the tapestries at their destination.

Princes going to the battle carried their tapestries with them to decorate war tents; for negotiations, or while receiving ambassadors. The war lord put tapestries behind him to impress the other party and with the subject matter well chosen—an idyllic love-scene for peace negotiations, a war scene for an ultimatum—set the mood for the conference. During intervals between battles, generals, passionate art collectors all, met privately to exchange and enjoy the collections of illuminated manuscripts and tapestries. If the battle was lost, or the prince captured, the tapestries were used to pay ransom. Jean, Duke of Nevers, paid Bajazet in Arras tapestries. Charles the Bold lost the battle at Morrat, and the Swiss took his tapestry collection as booty.

Tapestries were royal and ambassadorial gifts. King St. Louis sent a tapestry tent to the Khan of the Mongols. Tapestries were used at every festive occasion. The coronation of French kings traditionally took place in Rheims. The procession proceeded from the archbishop's palace to the cathedral between two walls of tapestries. In 1770 the city of Strasburg made preparations to receive the Queen Marie Antoinette. A raft was put on the Rhine, and an edifice built on this floating island adorned by a fabulous collection of tapestries. These were of different subject matter and Goethe was shocked to see tapestries of Medea among the religious subjects. This was considered bad luck.

Medieval Guilds. Medieval men were all members of the City of God. There was a corporate interwoven relationship of individuals, each man belonged to a group—a monastery, a guild, or the household of a prince. In the guilds, similar to the monasteries, members shared a common faith, common work, and had protection. It was very much like it is today in our country; the corporation, the club, or the labor unions, late offspring of guilds, unite groups of the population. The medieval guild had religious functions; each one had a patron-saint, they participated, as a group, in processions, arranged mystery plays. In peace the guild members ate, drank together; in danger they shared a certain point of the wall in the walled-in city and were responsible for its protection. Guild members took turns in services such as night watchmen, firemen, as burghers with a high sense of public duty.

The earliest record mentioning guilds is the Royal Charters of Worms (897-904). The first recorded guilds besides burial associations are the Weavers Guilds of Mains, year 1099. In Pavia a guild is recorded in the year 1010 and in St. Omer, France in 1050.

During the 13th century tapestry production was flourishing in Paris. According to a document—"Establissement des mestiers de Paris" dated 1258, there were both "tapissiers sarrasinois" and "nostrez"—our own. The name "sarrasinois" shows the Oriental origin of weaving techniques. In 1303 haute lisse, high warp looms were first mentioned. The rules prescribe: only good quality wool can be used; it is forbidden to work with candlelight; apprenticeship is eight years; and women are not allowed to work on tapestries. Silk weaving was established in Paris and the rules put down in the "Ordenence du mestier des ouvriers de drap de soye de Paris" dating 1260.

Arras, capital of Artois, was the center of spinning and wool trades. From England's green pastures came the wool which was piled up in storehouses on the piers along the canals in the low countries. Guild rules of the guild "Les Confreres" in Flanders, dating 1478-79 had strict regulations and penalties for violators. The weavers lived together like brothers, worked diligently.

Merchant guilds were first; craftsmen were permitted into the merchant guilds; then came the craftsmen guilds, although it is difficult to separate the two. A master craftsmen was an expert in his field and, by his authority, sold the product of younger, lesser craftsmen. The guild regulated conditions of sale, protected the customer from extortion, and the craftsmen from unfair competition, and the home market from wares from the outside.

The craft guilds assured high quality products by regulating materials to be used, set standards in workmanship, and with education and training, assured the continuity of high standards. There were heavy entrance fees to be paid and wandering craftsmen were collecting the money for years. With their wealth, the guilds built magnificent guild halls—their pride—and donated stained glass windows to the cathedrals, such as Chartres. The guild cup used for banquets was a treasure; the chest where documents were kept was a fine work of art. The Schwezerische Landesmuseum, Zurich, has fine samples of these. Registries of guild charters and official lists of members and other guild documents were painted on parchment with richly decorated initials. Giovanni Battista de Gavaleto painted the register of the Bologna guild of drapers in 1523. Guilds helped the members in legal cases, arranged tax reductions, military exemption, arranged marriages and gave doweries to orphaned daughters of guild members. There were no unemployed in the guild system.

During the 12th and the 13th centuries Paris was the center of art, culture, and education. Guilds of bookbinders, scribes, illuminators, were attached to the University of Paris. The School of Paris, great French illuminators, among them Jean Pucelle, created delicate and magnificent manuscripts—many a page is the finest painting ever done—for the clergy and princely patrons. In 1391 the first strictly French Brotherhood of Painters was formed, perhaps to prevent the French painters from the many artists from Flanders who worked in Paris, Angers, Dijon in the French tradition. In 1947 "L' Association des Peintres-Cartonniers de Tapisserie" was founded, its president was Jean Lurcat. However, since artists are individuals, this society no longer exists.

St. Luke was the patron saint of the painters guild. In Firenze the guild of painters, "Compagnia di St. Luca", had their rules fixed in 1478. Leonardo da Vinci was a member of the painters' guild in Firenze and owed the guild membership fees in 1472. Botticelli was behind with his payments from 1503-1505. Leonardo and later Michelangelo protested against the routine of workshops. Great artists and great innovators did not fit into the master and apprentice system. In 1563 the first real art academy, the Accademia del Disegno, was opened in Firenze under the patronage of the Grand Duke Cosimo I. de Medici. Art students became more independent and could work on their own. In 1571 it was declared that artists were free from obligations to the guilds. The "Academia dei Virtuosi", founded in Rome in 1543, was a religious institution, gave help to needy artists. Its patron saint was St. Joseph. Among its members were Bernini, Claude Lorrain, and Velasquez. Their annual exhibition took place on the porch of the Pantheon.

In Venice painters separated from the guild and organized their own society, the "Collegio dei Pittori" in 1682. The color formula of the "Painters' Manual of Mount Athos" was used until the 18th century. From the 16th to the 20th centuries in Europe, art academies were the centers where artists "belonged". Durer, the first great Renaissance man of the North, signed and dated his work; left extensive records and diaries. Interested in perspective and proportions, he had an independent, inquiring intellect.

Albrecht Durer (1471-1528) great German painter and graphic artist, of Hungarian origin, writes in his diary from Antwerp, 1520; "On Sunday, it was St. Oswald's day, the painters invited me to the hall of their guild, with my wife and maid. All their service was of silver, and they had other splendid ornaments and very sumptuous meats . . . So when we had spent a long and merry time together until late in the night, they accompanied us home with lanterns in great honor. And they begged me to be ever assured and confident of their good will, and promised that whatever I did they would be all-helpful to me. So I thanked them and laid me down to sleep."

The focus of life was the cathedral; activities centered around the square of the cathedral. The markets were all up in the Square. In one of his letters from Italy, Durer writes his mother, reminding her to take his prints to the market for sale. The highlight of the year for the artisans came when they presented their products in festive procession. Durer witnesses one and writes from Antwerp: "On the Sunday after our dear Lady's Assumption I saw the great procession from the Church of Our Lady at Antwerp, when the whole town of every craft and rank was assembled, each dressed in his best according to his rank. And all ranks and guilds had their signs, by which they might be known. In the intervals great costly pole-candles were borne, and their long old Frankish trumpets of silver. There were also many pipers and drummers in the German fashion. All the instruments were loudly and noisily blown and beaten . . . I saw the procession pass along the street, the people being arranged in rows. There were the goldsmiths, the painters, the masons, the broiderers, the sculptors, the joiners, the carpenters, the sailors, the fishermen, the butchers, the leatherers, the clothmakers, the bakers, the tailors, the shoemakers — indeed workmen of all kinds, and many craftsmen and dealers who work for their livelihood. Likewise the shopkeepers and merchants and all their assistants were there.

"In this procession very many delightful things were shown, most splendidly gotten up. Wagons were· drawn along with masques upon ships and other structures. Among them were the company of the prophets in their order and scenes from the New Testament, such as the Annunciation, the Three Holy Kings riding on great camels and on other rare beasts, very well arranged . . . At the end came a dragon, which St. Margaret and her maidens led by a girdle; she was especially beautiful. Behind her came St. George, with his squires, a very godly knight in armour . . . From beginning to end the procession lasted more than two hours before it had gone past our house. So many things were there that I could never write them all in a book, so I let it well alone . . ."

Durer was fascinated by ornate vehicles like the ones he saw in the procession at Antwerp. He designed a series of them in his woodcuts for the Triumphal Procession of Kaiser Maximilian; Victory leads the Imperial car pulled by richly draped horses. Fantastic animal creatures, ornaments like grotesque, allegorical figures fill the pages.

A procession of fanciful carriages promenade in various antique tapestries. In one called "Triumph of Chastity over Love", four winged white horses pull the ornate carriage of Love with torches, flaming altars, while Cupid is being dethroned by Chastity riding on a unicorn. Caesar and Cleopatra march along with many other allegorical figures in this tapestry which is crowded with symbols. This famous work was based on Petrarch's "I Triomphi" and its cartoons were probably designed by the School of Roger van der Weyden in Brussels. It was woven of wool and silk and is now in the Victoria and Albert Museum in London.

Late medieval tapestries with carriages were followed in the 17th and 18th centuries by tapestries with exotic themes, Orientalized carriages, camels, baldachins, and the romantic tapestries, showing gypsy caravans.

The automobile and the airplane have not yet been the subject of tapestries, but speed and motion which shape the contemporary scene are reflected in tapestry design. Artists are also inspired by the scenery viewed from airplanes: the clouds, sun, the earth or the sea. Planets, stars have moved the imagination of tapestry artists, like Domjan in his latest cosmic series.

Apprentices in the guild system. Labor, manual work was despised in slave societies like Greece, and Rome. Nor did the leisure classes in the Middle Ages work. The knights engaged only in battles or in training for warfare and their pleasure was in hunting and games. All of this changed slowly during the late Middle Ages, as guild members became wealthy, powerful citizens, whose sons inherited their privileges.

The life of the apprentices depended on individual masters and even on their wives. Large workshops had half a dozen young boys who were learning the trade over several years. Their apprentices were members of the household, ate at the same table, joined in family prayer and shared domestic duties; they served at the table, chopped the wood, carried in water from the well. The common hall was turned into a dormitory at night. One can imagine the boys trying to avoid hard chores and punishments.

In good weather much of the work was done in the open; spinning on the doorsteps, washing the wool in the river nearby, dyeing the wool in the backyard. Working hours were long — no daylight hours were wasted.

By the time the training of an apprentice came to an end, the young man had finished a masterpiece of craftsmanship. After this he usually embarked on a journey — "Wander Jahre" — often to traveling foreign countries to see and learn more.

* * *

Cennino Cennini writes in 1400: "Begin to submit yourself to the direction of a master for instruction as early as you can and do not leave the master until you have to." Fine and applied arts, arts and crafts were not yet separated. The

37

original meaning of words like *ars* and *arte* was technique or craft. Many of the great painters and sculptors of the Renaissance started as goldsmith s apprentices. It was fundamental that they learn the use of tools, instruments, and study proportions, the properties of art materials, and the right way to handle them. A wealth of knowledge was acquired from a good master while he collaborated with his students and apprentices, giving them more and more responsibility in the work. Learning guarded formulas for mixing colors and a philosophy of life while helping and imitating the master, the student apprentice learned and prepared himself for an independent career. There were no books or schools where he could obtain this knowledge; tradition was handed down from generation to generation and new inventions were made in the workshops.

The Master Weaver. The master weaver, *chef d'atelier,* supervised the work of the weavers, *tapissiers;* under them worked the apprentices. The master also handled the shop, storage and counting house. He had to negotiate with the patron — even the king — as tapestries were not usually made for the average man.

The master weaver directed the artist cartoonist or changed the cartoons, the size, the borders according to the patron's wishes. He also selected the materials, wool, silk, gold and silver thread, and calculated the quantities needed for each shade. The dyeing of the wool was of the utmost importance; some dyes were very expensive; the brightness of color, the luminosity, and durability were all the concern of the master weaver.

In weaving, fine detail was used only on faces, jewelry, a few accents while the rest was simplified according to judgment. The "battage" system of parallel woven lines by which color transitions were achieved was greatly developed. There were strict rules: long narrow parallels were used on drapery, short lines for leaves, feathers of birds, shortest battage for the ground, earth, rock. Usually contrasting colors were used. To compensate for the small number of colors available, the master fully employed battage, produced shadings, modeling.

The master weaver often had no cartoon but worked from his own sketch, enlarged figures or motives, or rearranged them. Woodcut prints, or a page from an illuminated manuscript may have served as samples. The builders of the cathedrals had their collection of ornaments, motifs used for capitals, plants, vines, animals, monsters; the collection was traveling with the masons, similarly the weavers had their sample book. Until the turn of the century there was a general belief that builders of the cathedrals, weavers of tapestries were unknown; research in guild books and other documents, however, has changed this concept. Some seventy masters weavers have been identified through deciphering signatures, initials, lettering, or little signals. Means of identifying others were lost because tapestries were often cut and the identification marks were on the edges, as a rule. Some data has been preservd on the life and work of the weavers and cartoon painters we know by name.

Nature in Tapestries

Forest *Dongan*

"Les Milles Fleurs." Looking at a "milles fleurs" tapestry one feels the joy of a flowering meadow. The eye, like a butterfly, flies from one flower to the other, enjoying the intricate details of plants and animals.

Gothic tapestry is two-dimensional in feeling. The figure in early Gothic tapestry stands separated against a plain, unbroken background. The void is filled later with flowers, birds, plant and animal motifs freely scattered around from border to border; this is the "milles fleurs" tapestry. There is no sky and no horizon, the small individual bunches of plants stand clearly against the dark green-blue background. There are also tapestries without figures — mostly heraldic tapestries, coats of arms with "milles fleurs" or "verdures" background. The "verdures" can be stylized leaf scrolls or more realistic branches of leaves, from small life-size leaves to enormous ones. The "milles fleurs" style was very convenient for changing the size of the tapestry according to order — the master weaver had the cartoon of the figure and filled the required space with plant, grass, flowers, bird motifs from his sketch book.

The origin of the "milles fleurs" is not known; it is possible that it derived from the religious custom to fix bunches of flowers to a linen cloth and hang it out of a window as a decoration during the Corpus Christi procession. The bunches of flowers were woven into the tapestry and the tapestry hung for decoration during the procession. Tapestried flowers did not wilt.

The Domjan flower tapestries show a deep love and understanding of nature — drawn in a Western way, but with a touch of the East. His color woodcuts of botanical subjects were called flower portraits. They were done in a large monumental size and are the dream of a garden, a subconscious world where dimensions change, disappear; the petals envelop, surround, the green colors sing, the foliage recalls the forest, the smell of moss, and fern, the mysterious deep dark places of the untouched forest; the flower petals in contrast reflect the sunshine, open radiance. Surrounded by the Domjan flower tapestries one feels like a small bird or beetle. The flowers are very large, fill the whole tapestry from edge to edge in a rich composition of brilliant colors.

The White Poppy is a large pale and light flower with open petals on the upper left side of the tapestry facing the viewer. Petals are curling, folding, and each has a deep velvety patch at the base. The center of the flower is green with white pestils. The white of the pestils is, however, a very light green, and the black patches are of four colors — a very dark purple, brown, a spot of pale crimson, an echo from the petals. The dark spots, about six inches in diameter, are of four shades. The petals are much larger forms and, of course, many more colors are used; basic colors are a pale pearl grey in several shades and a very light crimson. Then there is a touch of cadmium yellow like the gentle caress of the sunshine. With hachures, battages — methods of shading — transitions of the colors, the petals are soft, delicate, and very colorful; their effect is achieved by a fine modeling of close shades and tones. A second flower to the right with folded petals is kept in darker shades of rose-mauve, a bit of yellow on the contour lines and the black spots, which in this case are of a very dark English red, a grey and a green. Stems are of ochre, brown, and green and there is a charming bud just opening; and two flower pods with the last petal just about to fall off — all elements of a perfect botanical representation, yet the tapestry is pure color and poetry. The lower part of the tapestry is a luscious verdure, green leaves, almost sensual, strongly shaded, three-dimensional, with strong, accentuated veins. Combinations of five green and brown colors are used for this area. *The White Poppy* tapestry has twenty-eight colors; wool; no border.

Queen Anne's Lace. This tapestry by Domjan is a large open flower, en face, two-dimensional as one would prepare it flat for pressing; the flower dominates the tapestry. There is a smaller blossom in blue and two closed ones in reds, with stems, their narrow leaves intercrossing. The composition is built up so that the eye always returns to the cadmium red small dot in the center of the large flower. A Queen Anne's Lace flower in the field is a delicate, pale off-white lacy flower, always seen against the green meadow; and so that is how it appears on the tapestry. A more careful look reveals the colors of yellow in several shades, rose, mauve, blue; there is not a single white spot anywhere. Not a bit of white wool was used, but the effect is that of off-whites. Green, red and purple are used in the leaves.

The lovely lace effect comes from so many broken up shapes and forms within the individual segments of the big flower. One segment is broken up by straight lines at right angles. Each little square has a dot, a triangle, a smaller square. Another segment consists of rhythmically repeated bigger and smaller dots; one has stripes, the other little floral forms, in so many variations, positive and negative forms and shapes. In this the Domjan tapestry is unique. Not the hachures, battages, or piques give the variations but small individual color forms. This is Domjan's very own device by which his tapestries can be identified. There isn't anything else like it in the history of tapestries.

The background of *Queen Anne's Lace* is of several shades of green and has the effect of moving, like the green shadow of grass on a morning meadow. There is a contrast of green and red, light and dark. This tapestry has no large clearly defined decorative color shapes. It is all broken up into countless small fine details from edge to edge in endless variation. It stands out as a very rare and unusual design.

The Big Poppy. If one flower had to represent all floral art, this would be it. There are two flowers, the large one at its height of splendor, with ruffled petals, in purpled ruby; the calix forms a dark shape in the center. The second flower still closed in a darker English red. The three large scrolling zig-zag edged leaves recall curved acanthus motifs.

The battage, hachure, is almost invisible with many delicate fine color transitions, also in the green background. There are no small broken up shapes, the forms are clear and decorative. The Big Poppy is grandiose in its simplicity. It is a strong voice; it can be mixed into any gallery of all kinds of art work and will come out victorious. This is the tapestry that can create a garden on a bare wall.

The Garden and Plants in Tapestries. A loving couple is sitting in a garden, arms interwoven. The man holds the lady's breast tenderly and talks to her, looking at her with expectancy. In front of them is a sextagonal stone table, tilted two-dimensionally, with a blue striped table cloth, a wine decanter and chicken on a plate. A young man, with a smile on his face and a garland of wild flowers in his hair, cuts a bunch of grapes from a grape vine growing on a trellis. The grapes are beautifully observed and drawn, with leaves and tendrils. A bird behind helps himself to the grapes. A thorny braided fence of acacia branches below indicates that we are in an enclosed garden. Violets and columbines grow in the grass. This is one of several scenes of the *Liebesgarten Mienneteppich,* a tapestry, woven in Basel, 1460-70, of wool, linen, and silver thread (102 x 355 cm.)

Groups of people are gathered in a garden for the spring festival on the tapestry entitled *La Primavera.* Two flower-festooned arbors, diminishing in perspective left and right, lead to a stone arch in the background. Cypress trees, pale in the distance, add to the three-dimensional effect. To the right are musicians with wind instruments, to the left part of a villa is seen; the owners just stepped out to the terrace to receive a group of girls. In the foreground children with flowers watch the festival. The tapestry, made in 1640-43 in the Medici workshops in Florence by the weaver, Bernardino van Asselt, has a rich bordure of festoons of fruit and plants; urns and architectural elements.

Floralia was an ancient spring festival when young men went out to the country and returned with green branches. Medieval pageants of May followed. In the illustrated manuscript of *Les Trés Riches Heures* of Jean, Duke of Berry, page of May we see a group of elegantly dressed riders following a group of musicians to the tune of flutes and trumpets. On this day the King distributed green robes known as *livrée de mai.* The green color used in the miniature comes from malachite, a crystal of a delicate soft pale green. Three girls are dressed in these green robes; their horses are also caparisoned in green and gold. One nobleman is dressed in an azure-blue gown, richly embroidered in gold. Another wear the red-black-white royal livery; he is probably a prince and the sumptuously dressed girls are probably princesses. All have garlands around their heads or necks. There is a flowering shrub in front and the forest behind with a chateau against the sky.

More than one hundred years after the Duke of Berry's May pageant, Sandro Botticelli painted the *LA PRIMAVERA* (1477-78), magnificent nature study. This allegory had classic elements; the beauty of the human figure in dance movements, the three graces; poetic symbolism is in Rennaissance style, but the bunches of flowers on the flower-decked girl, the flowers in the grass in small bunches; carnations, forget-me-nots, daisies, the lacy branches of apple trees, each leaf standing alone, are very reminiscent of the medieval "milles fleurs" style.

The word, Paradise, comes from the Persian, *pairidaza.* Its original meaning was *wall, enclosure;* later it meant *the enclosed; the park, royal gardens.* The picture of the enchanted garden of Eden, the everlasting dream of human happiness, the picture of the oneness of man and nature has been with us throughout history.

Plants are used for their symbolic meaning in art and in tapestries; lilies stand for purity, innocence; red roses and thorns are associated with Christ's sufferings and Calvary.

In Byzantinum flowers were identified with stars. The azure skies of the mausoleum vaults of the Galla Placida mosaics in Ravenna are strewn with identical white flowers for stars.

Flowers, plants, are readily available, easy to draw. Small wonder the "exempla" had many floral motifs, not all taken from nature. Flowers for their

variety of shapes and color, leaves, branches, trees for their rhythmic qualities and all botanical subjects for their happy associations were used more often than anything else in illuminated manuscript borders and in tapestry backgrounds, bordures.

Herbaria. The collection of plant drawings put together was called the herbaria. Drawings, water colors, woodcuts of botanical subjects presented the results of scientific study of nature. The complete plant was drawn with roots and fruit in the *Medici Herbarium* a collection of water colors dated 1570 and done by Giacomo Ligozzi. Even plant diseases were included.

Islamic illuminated manuscripts dealt with plants even earlier. *The Book of Antidotes,* dated 1199 and made in Northern Iraq shows men hoeing, watering, cultivating plants that were used as medical herbs. Their healing qualities are explained in fine calligraphy next to the illustrations. In the *Materia Medica* of 1229, a doctor teaches his disciple the use of a plant as medicine. On another page of this manuscript one plant fills the page; leaves, branches, stem, root system, soil-line are clearly understood and drawn. This is great progress. An Armenian Gospel illustration 900 years earlier uses bunches of grapes — Abraham and the Angel sit under a trellis of grapes. Grapes and leaves are drawn separate in the manner of a textile design — there is no stem to connect them; the illuminator did not understand the organic structure of the grape plant.

In later periods realistically recognizable flowers were made to perfection in Islamic miniatures. Prince Humay arrives and is received by the Princess Humayun, his sister, and her attendants, in China, in the palace garden of the Emperor. The figures of this illuminated manuscript page stand separate, without perspective. The charming and gracious girls, beribboned in a flowery background, look very much like the "milles fleurs" tapestries. Flowering almonds, cherry tree bloom, iris border the garden. Bunches of flowers, with buds, leaves, fire lilies, daisies, poppies, ornamental bushes are well observed. The miniature by Herat (1291-1352) made for Baysunghur shows an enclosure, closed-in-garden, outside a night sky, while there is daylight in the garden (an early touch of surrealizm?).

Drawings in mediaeval herbals, like bestiaries, were often copies from earlier sources, with mistakes and misunderstandings included. The invention of the printing press did not help this situation; rather the contrary. Printers wanted to keep down the prices of the books and used and reused existing woodcuts. They even inserted woodcuts that were associated in no way with the text. One schematic plant had to stand for several different species to provide the book with illustration in sufficient quantity. A printed, woodcut-illustrated book which was a herbarium from Monte Cassino was published in Rome in 1481.

Gart der Gesundheit, a book of herbs, was published in Mainz, in 1485; and *Hortus Sanitatis* was published also in Mainz in 1491. The *Neu Kreuterbuch* was published in Basel, 1542. In this case the writer-editor asked for complete pictures of bud, seed, and developed plant; the joined effort of artist, who draw the plants, woodcutter, who prepared the book for printing, cut the blocks, and the printer, who printed it, the writer, who made the text and the editor, resulted in the first true set of botanical illustrations. The Hunt Botanical Library, in Pittsburgh has one of the largest collections of botanical illustrations from incunabula to contemporary artists including Domjan.

45

Stone Age Man. We know that born in every child is the love of flowers, but that Neanderthal man appreciated flowers and had the patience and skill to weave them into garlands. These amazing discoveries were made just recently, and may change the present mistaken picture of stone age man. Soil sample analysis from the Shanidar excavation of Neanderthal skulls and bones showed clusters of flower pollen. More than a dozen different colorful small flowers were woven into branches of a pine-like shrub. From archeological finds the story was reconstructed; a victim of an earth slide was buried with flower garlands 60,000 years ago, certainly proving the creative interest and artistic talent of early man.

Plant Motifs; Plant forms were used in Egypt, bundled papyrus columns in the Tell el Amarna tombs, 1360 B.C. Columns often end in a capital of a bunch of stylized palm leaves — palmiform — curving outward in the temple of Thutmosis III., rebuilt by Amenhotep III., 18th dynasty. More naturalistic are the palmiform columns' of the Ptolomaic period. Half-open lotus blooms emerge from palm leaves. Palms, lotus, reed, flowers, buds, were used in architecture and bas-reliefs from 800 B.C. these floral elements were used in rich interlacing and the decorations painted in bright colors of orange, purple, brown. These same floral elements appear as woven ornament on textiles.

Assyrian royal palaces were decorated with palmette motifs and later the pomegranate; the life tree of Assyria. The pomegranate was a fertility symbol of Rome. In coptic tapestries it was connected with the suffering of Christ, and continued its role up to the Madonna of the Pomegranate by Botticelli 1487. Etruscan tomb paintings are rich with plant motifs. Figures dance, perform religious ceremonies and between the figures trees are painted, branches, leaves, fruit, not in a naturalistic way, but handled as ornament, ondulating lines, with an easy routine. In the Knosos Throne Room stylized griffins face each other in a lily garden. Minoan vases are decorated with lilies, anemones.

Greek floral motifs include olives, laurel, grape vines. The Greeks used laurel — laurus nobilis — an ancient medical herb, in garlands to honor poets, victorious athletes and war heroes. Among ancient medical herbs the ancient Greeks also used Borage, the herb that makes men merry.

Paeon was the God of Healing, and his plant, the peony, considered a magic plant. The Romans also used laurel and this tradition continues during the Renaissance and is still with us in symbolical decorations of diplomas, prizes, etc. The Romans were extremely good at making garlands of olives, oak, laurel, bunches of roses, fruits, grapes, ears of wheat tied with ribbons and they decorated altars, columns at festivities, banquets, used them for table decoration. The classical garlands of fruits and flowers were extensively used in the Renaissance. Wall-paintings, stucco and stone-carved decorations of buildings favored classical garlands. The beribboned fruit and flower garlands became a standard for tapestry bordures since the Renaissance.

The *acanthus* was a pre-Hellenic magic herb of eternal life. The ancient Greeks used the acanthus motif on funerary steles. According to Vitruvious, the Corinthian sculptor, Kallimachos made the first Corinthian acanthus capital, end 500 B.C., after he saw a basket of acanthus leaves on the tomb of a girl. The acanthus leaf was the floral motif that, in the form of the capital, existed for over a millenium in infinite variations; ancient Rome made the most of it.

On Romanesque capitals it took surprising forms; during Renaissance neo class-icism the Corinthian capital flourished.

Bronze railings of the Carolingian Palatine Chapel of Aachen have a round eyelet out of which the acanthus leaves emerge. Illuminated manuscripts are bordered with acanthus and the eyelets grew into elaborate designs. Whole pages are filled with gracefully curved acanthus leaves, the reverse side in con-trasting colors such as blue with red, gold with green, drawn delicately and shaded like curving, folded ribbons, in the scriptoriums by illuminator — cali-graphers who never saw an acanthus plant, nor did they know the origin of these leaf forms.

A buck and a lion, a deer and a unicorn, all decorated with bells, face each other. Large curling acanthus leaves in blue-green, shaded in yellow and red, crossing each other, form the background and fill the tapestry, *"Teppich mit Symbolen Fabel tieren in Blumenranken"*, wool (114 x 185 cm), woven in Basel, 1480. At this time the acanthus motif was widely used in tapestries but its use goes back to much earlier times, in fact, to Coptic tapestry.

Acanthus leaves, even larger in size than in the Basel tapestry, fill a verdure tapestry of Aubusson titled *"Bêtes et feuillages"*, (16th century, Fougerol col-lection). The rather schematic lobated acanthus leaves are woven in six shades of a dull blue-green wool. Between foliage dogs attack a bull, there are birds and a deer; farmhouses and trees in the background and a simple monotone bordure of bunches of apples and leaves.

On the Hagia Sophia repetitive acanthus motives are spiraling, with palm-ettes, rosettes, vines. The cross appears superimposed on a Sassanian palm-life-tree.

In India the *lotus* blooms; Buddhas and Botthisatvas meditate on lotus thrones. Buddha was born under a tree, reached enlightment and was preaching under a tree. Trees, branches, leaves, fruit, are elements used in Buddhist art. In India women wear fresh flowers every day. Lotus, all sorts of flowers and petals are sold around temples and taken into the temple as religious sacrifice. Even for the poor, flowers are an everyday necessity.

Flower-decorated gods, symbols, carriages, have an important role in fes-tivities especially in the South where these traditions are kept very much alive. Bananas, phallic symbols are used for fertility festivals. Ancient Hindu archi-tecture, temple towers appear as buds, gigantic stone flowers, organic forms. Acanthus leaves reached India and with Hellenistic motifs appear in Gandharan art.

Islamic flowers are roses; rose-oil, rose perfume, rose honey, rose-water scent-ed candies are some of the Turkish delights; also carnations, lilacs, hyacinths, and tulips. Tulip bulbs were guarded treasures of Sultan's gardens; they were stolen and smuggled to Holland. Oriental carpets and tapestries use these flow-ers, carnations, sun flowers, tulips combined with the palm-cyprus life tree. Seed pods are also a motif of rugs and tapestries.

The most typical Chinese plant motiff is the bamboo, subject of thousands of poems. It is the symbol of friendship, and of the gentleman; it is flexible, bends in a storm and stands up again. The Bamboo School of Painting for brush drawings taught how to paint bamboos for scroll paintings and screens.

The pomegranate in China is the symbol of family blessings, children; like the plum, it is a symbol of spring, new life. The pine is the symbol of longevity. Peonies are the symbol of wealth and the orchid of feminity. The lotus is the symbol of purity, from muddy darkness, it finds its way to the surface of the waters and to the sun.

In tapestries as well as carpets and paintings, gardens are frequently used for their beauty, or for their symbolic meaning; the garden of love, the garden of virtue, the garden of delight, and *Paradise,* the garden of Eden. The garden is either a subject by itself or the background for a scene. A garden is to guard, surround, protect, a garden is a limited space in which nature can be enjoyed undisturbed; a garden is a miniature universe.

While bloody battles are recorded in history often by the hour, there are no records of history about hours spent happily in a garden; yet there were gardens to enjoy. On a tomb painting in Thebes, 1400 B.C., there is a garden; a rectangular pool in the center holds budding and blooming water lilies; fish and ducks are swimming in the basin. The artist of ancient Egypt could not express the difference between the fish swimming in the water and the ducks and lotuses floating on the surface. Around the basin are flowers and trees planted in a row; palm trees, fruit trees, bushes. The conception of the painting is two-dimensional, trees at left and right are drawn horizontally; what is behind is placed higher up, there is no perspective. Islamic carpets two thousand years later used the same two-dimensional representation.

Egyptians used flowers on many festive occasions; the blue lotus of the Nile, the white lotus, jasmin, oleander, acacia, tamarisk, and fig trees were grown. There were arbors of grape vines and trees provided relaxing shady places. In the Metropolitan Museum of Art is a small wooden model house and garden from the tomb of Meket-ra, Thebes, II dynasty. The house has a covered porch with columns that opens to an orchard, surrounded by a high wall. The model shows the perfect rendering of tree structure and leaf system.

Abu was the lord of plants, the Tell Asmar sanctuary was consecrated to Abu.

Sassanian emperors had luxurious gardens. To compensate for them in the winter large floral carpets were laid on the floor. Persian gardens were luscious with flowers and various subtropical plants. Plantings were made in geometrical order. There were large areas for the ceremonial royal hunt; these were forests and meadows; a man-controlled and improved wilderness. Successors of Alexander the Great learned gardening from the Persians. The Greeks had no gardens but temples had orchards planted around them, and there were the sacred groves.

One of the seven wonders of antiquity were the suspended, terraced gardens of Semiramis, Queen of Assyria, a lengendary figure later associated with the goddess Istar-Astarte, who erected great monuments and built the banks of the Euphrates, probably for irrigation. Her victories over the Medes and the Chaldeans are long forgotten, the monuments crumbled, but the legend of her gardens still move our imagination; the gardens were probably a series of roof gardens, terraces, and balconies resting on heavy arches, an artificial mountain or terraced pyramid on the flat land. They were built about 800 B.C.

In ancient Rome both the oriental influence and the Dionysus cult brought forth gardens. The rock garden derives from the Dionysian ceremonies, the background of which was the sacred landscape, inhabited by the demi-gods.

Private pleasure interplayed with religious overtones in the Roman garden. On the right bank of the Tiberis Caesar built large parks, Horti Caesaris. Maecenas, Nero, and others have built parks and donated them to the citizens of Rome. In 55 B.C. the first promenade — the corso through public gardens was created. It lead to the theater of Pompeii with trees, shaded avenues, and fountains. Tree-lined avenues were leading to the Thermae. Cicero called landscape architecture "ars topiaris," a Greek term, meaning not the trimming of trees, but the design of the landscape, choice and arrangement of trees. Even in these early times there were ground covers, ivy, periwinkle to cover walls, grottoes, fern was planted in the shade and cyprus trees along the roads. The selection of plants, planting cultivation is agriculture when they are useful and horticulture when they are planted for pleasure and beauty. When the garden becomes an extension of the house with porticoes, pavilions, steps, terraces built, it is then the field of the architect or landscape architect.

In the house of Neptune and Amphitrite in Herculanum, the Nympheum has a mosaic decoration with peacocks perched on garlands of laurel and grape vines grown from an urn.

For centuries, tombs were robbed regularly. Then came scientific archeology. The outlook of archeology has changed completely during the last 75 years. Earlier the motivation was to dig for a precious object, a gold treasure, which, when found, was immediately removed from the site. Today the aim is to leave finds "in situ." A new branch of archeology, pale obotany studies the negative imprints of roots in the soil. This was done in Pompeii. It is possible to tell what kind of plants were growing, the age of a fruit tree in Pompeii, and gardens can be replanted. Pear trees are bearing again in Pompeii as they did 1900 years ago, with identical irrigation restored. Not a gold object, but a small hollow space in the ground is the find; precious information to broaden our knowledge of times past.

Paradise. Islamic warriors faced death undaunted; in moments of danger, in their minds emerged the picture of *PARADISE:* a marvelous garden with cool fountains and sweet girls the reward promised to heroes who died for Islam. And with this picture from a small group of fanatic warriors starting from the Arabian peninsula, there became a world empire from Granada to Agra, from Istanbul to Budapest. Much of the conquered territory was desert land but gardens were built nevertheless. Walls protected the gardens from desert sand storms. Inside, the main concern was water; a reservoir was built, or water came from wells, there was always a canal and irrigation system, to water the trees and flower beds. The warm climate and water produced flowers of exceptional brilliancy. On uneven terrain Islamic gardens were built on terraces; the water flowing down in cascades of waterfalls. Colored glass oil lamps illuminating the waterfalls at night must have been stunning.

Sometimes the water was directed into narrow pipes coming down-hill onto the fountain where the water pressure produced jets spouting in a planned design. In an Islamic illuminated manuscript a drawing shows the water works, great ingenuity, mathematics, engineering knowledge was needed before it came to artistic effects. If needed, water towers were built and hidden behind garden pavillions, often there was a pool in the middle of the garden, rivulets flowing into it or out of it in four directions, dividing the garden in four parts. River banks, built higher than the land, had cyprus trees, grapes.

Hindu Maharajas also built luxurious gardens. On an Indian miniature the Islamic influence is evident. A loving pair walks in the garden holding hands. Behind them is a garden pavillion of sextagonal shape of white marble inlaid with floral motifs in green and red. Large shade trees with climbing vines and flowering branches surround the pleasure pavillion, which is provided with pillows for sleeping and shades that can be pulled down. The pavillion is on an island, rosy lotuses are blooming on the water and a pair of ducks is seen in the corner. The lovers are bedecked with rich jewelry and the lady wears a translucent sari with rich gold border. The gold leaf work is very delicate and the painting from an illuminated manuscript shows the feeling for tender details and also that divine state of mind; the oneness of man, beast, and nature in the created world.

The Cloister Garden. On a medieval panel painting by the Master of the Middle Rhine, 1420, titled *"Garden of Paradise"* we see a walled garden "hortus enclosus." The Madonna is reading a book. Sitting in the grass, an angel is in lively conversation with two men under a tree. The Christ Child plays with the angel's harp. A girl gives water to a bird perched on the rim of a well; another girl is harvesting cherries in a basket. Lily of the valley, lilies, iris, thistle, and wild strawberry plants are growing in the grass. Gone are the Byzantine golden backgrounds, decorative palm trees; the sky is blue, and inhabited with birds. The earth is green, the plants depicted could grow anywhere. All pleasures are here; a lovely garden, music, literature, good company, and fruits to eat; on a sextagonal table the left-over peelings show, the fruit has been tasted. Cherries and carnations, for their red colors, represented the Madonna and Child in icons, and referred to the Passion. There are probably many symbolical meanings in this painting, elusive for us today.

Benedictine monks started agriculture in the fertile Po valley. Vegetable gardens and orchards were planted and the monasteries were teaching the people how to plant and care for gardens. Within large monastic building complexes was the cloister garden for religious meditation and prayer. Vaulted and arched walks were surrounded by high walls. The secluded, protected cloister garden had a fountain in the center, a few fruit trees were growing, flowers, lilies for the altar, perhaps herbs. Secular gardens were made on the same principle with arches around and a fountain in the center, especially Mediterranean countries continued this scheme through the Renaissance.

Cloister gardens whose silence is part of them, silence and church bells ringing, also birds singing, filling the space; lonely steps seem to echo through the corridors or is it our imagination since the nuns, and monks are gone? The sunshine passing from one richly carved capital to the other, illuminating in turn interlacing, scrolling floral motifs, geometric ornaments, strange beasts; while a shadow play of arches falls across the corridor. The Basel Munster cloister garden with delicate stone carvings, a lace of ribs, leafs, interwoven almost oriental in effect. Orange trees bloom in a sunny corner of the cloistered garden in Segovia, through the holes, the negative pattern of arches there is a sharp blue sky. The moss covers the fountain and blue, orange, yellow ceramic tiles give a brightness that contrasts with the forbidding walls in the Hidden Nunnery of Pueblo, Mexico.

Imaginary buildings, building elements, a silhouette of lacy archways were often used in tapestries like the *Nine Heroes* tapestry in the Cloisters, New York, with figures framed in arches, in a fanciful interwoven design.

Cloister gardens remained almost intact because of their small size, and built-in situation, not much could be added in following epochs. Larger and more open gardens built during the early Renaissance underwent frequent changes, complete remodeling with the changing fashion of subsequent styles, mannerist, baroque, rococo. A few descriptions in literature and some art works give information about gardens gone. There is no garden museum of reconstructed gardens, a history of landscape architecture. Plants grow, and it is difficult to keep a garden at a certain stage. Who knows how long it took for the tall, trimmed avenues to grow up in Versailles? How tall were the trees during the Sun King's life time?

Castel Nuovo in Napoli, this ancient fortress against pirates of the bay with a magnificent view of Vesuvio and, on clear days, of Capri in the distance, had buildings and gardens added during the Anjou King Charles. In the Decameron there is a description by Boccaccio of a garden; the garden is laid on a geometric system of walks with rectangular and triangular flower beds of roses and other fragrant flowers. The walks are bordered with jasmine bushes. Vines are grown on rustic arbors. Laurel, cedar, olive trees are planted. The center is a space of closely cut grass, bordered with flowers. This is the first time grass is used in a garden; before it belonged to the pasture and meadow. The 14th century garden also has statues, fountains, flowers are the iris, lilies, roses, violets from medieval times, now in a more symmetrical formal arrangement, and the repertoire is widening with the crusaders, merchants bringing plants, seeds from the East. Gardens include aviaries, nets enclosing whole trees where birds could fly and nest.

Renaissance gardens influenced by the antique revival were designed by great artists, architects, like Alberti. Garden planning became a highly specialized work. Man at this time felt he was the ruler of the world, and wanted to organize, tame, and reshape nature. Chance hills and rocks were not allowed, the land was evened out, with shovels. The garden was calculated with a ruler in geometric order on a large scale. "Opus topiarum," shaping of shrubs and trees had a modest start during the Renaissance; later it was carried to excess. The box shrub was most tolerant to pruning, trimming; it was used to border flower beds. Later most intricate shapes, ornamental patterns, even labyrinths were made of trimmed boxwood. The box shrub was also used as specimen spheres, cubes, squares, romboids were formed, often superimposed. There were urns, and vases of box shrub and even animal shapes were trimmed.

Michelozzo was the designer of the Medici villas. Raphael designed gardens for the Medici Pope, Gulio II while he was cardinal in Rome. Raphael first included a labyrinth that became most popular and was later imitated. Bramante started the Vatican gardens in 1505 and built the three terraces between the Vatican Palace and Belvedere; later the library, loggia were added and the gardens rebuilt. The Pitti Palace by Bruneleschi in Florence has the Boboli gardens in the background. The gardens can be seen on a sixteenth century drawing and show the principle of organized nature; like Italian gardens of the period, they are built on a central axis with right angle avenues that divide the gardens

51

Nephertiti in the Garden Domjan

into exact symmetrical parcels of orchards. The fountain with statue is in the center and the trimmed lawn. Straight lines were enjoyed and the perspective. Buontalenti made the grotto in 1583-88 with artificial rocks and groups of statues. According to Bramante the garden has to be built on a well defined center line, this was followed with walks, plants arranged in military order — the fantasy, mystery, the need for disorder found an outlet in the grottoes. Raphael emphasized the importance of the main building in relation to the surrounding garden. Palladio designed avenues of trimmed trees. During the 16th century the clear open space became more complex. There were many private villas built with grandiose gardens around them, in Rome, and elsewhere. The gardens have always been as they are today a pleasure in the lovely climate of Italy.

The *Great Unicorn Grotto* wall of the Palazzo Borromeo on Isola Bella, Laggo Maggiore, seventeenth century, is a system of artificial grottoes, niches, with shells, sculptures, balconies, needles, pillars, or several stories; on top is the unicorn. Italian gardens were copied in Europe; Les Jardins de Luxemburg in Paris were made for Queen Marie de Medici. In Buda the magnificent terraced gardens with marble fountains, exotic birds in golden cages, bronze statues, loggias, paved with faience tiles bought from Italy were made for the legendary King Mathias Corvinus and Beatrice de Aragon of Naples.

Gardens of Versailles; Italian gardens were first followed than surpassed by the French. The designer of Versailles gardens was André Le Nôtre, born in 1613 in a family of gardeners. He studied architecture with Mansart. At the age of 24 Louis XIV made him superintendent of the royal buildings. French gardens and parks around the chateux with civilized trimmed trees, straight avenues and rondoes, fountains, urns and statues, created a refined and artificial scene as opposed to the natural landscape of the countryside. Principles of Italian baroque gardens were followed but with French elegance, and a great taste in elaborate details. Walking in a well groomed park bright with flowers, sitting on a chair under trimmed trees is a way of life in France. The many different small and large parks made city life enjoyable in Paris; around the chateaux they are inseparable from the building. The countryside of Ille de France is amazing, miles after miles of espalier trained orchards, vineyards — these need years of practiced skill — and perseverance.

Le Nôtre, when he laid out the plan for the gardens of Versailles concentrated on two main views; the view of the chateau from the garden and the view of the garden from the chateau; no trees should block the view, instead, intricate, symetrical low flower beds were planted. Inside the palace one could overlook the vast expense of gardens: the open-lawn areas, "Tapis vert" and in contrast the straight, cool, shady avenues with tall trimmed trees, and the diversity of curving side paths with the trimmed box shrub. The King's favorite were the orange trees grown in silver tubs in the *Great Hall of Mirrors.*

Repeated rondoes are most typical of French parks. The gardens were added to and became more complex. To escape from the large formal buildings and gardens Marie Antoinette had the charming, intimate *Le Hameau* built, her toy farm-house where she could play shepherdess.

Schonbrun Castle of the Habsburg Emperors was made after Versailles, with terraces, stairways between the Castle and *Gloriette,* fountains, gold fish.

53

Trimmed alley's leading into shady forests and open areas in the center with formal flower plantings.

In the summer palace of the Bourbon Kings of Naples, in Caserta, parks and fountains designed by Vanvitelli became monumental hydraulic works. The water is flowing down from a hill in a straight line interrupted by basins, cascades and waterfalls. Ceres, Diana, and a group of deer adorn the *Basin* of the *great cascade.*

A reaction to the geometric formal French garden was the English garden of the 18th century. A back-to-nature movement was combined with romantic spirit. The idea was to enjoy the virgin beauty of nature. This did not exclude elaborate preliminary ground work, including artificial ponds, these, however were irregular in shape, imitating nature. Under the shadow of untrimmed trees, willows were placed ruins. A narrow, winding path lead here unexpectedly. Antiques were cherished and English gentlemen home from the "Ground Tour" wanted to have Roman ruins in their garden, a broken column, a whole temple; artificial, if the real one could not be had. There was not much color. Carpets of well kept green lawn is the basic element of English gardens; the climate cooperates. Later it became fashionable to build Chinese pagodas.

English gardeners got plant species from distant lands, the Americas, China, India. Many were lost on the way or later; other were acclimatized. Scientific interest resulted in arboratums, botanical gardens, and palm gardens.

Style nouveau and Pre-Raphaelites were using plant motifs to design wall papers, lamps, screens, even furniture with ondulating stems, branches, leaves. A tapestry woven at Merton Abbey, designed by William Morris (1834-1896) is an illustration of his poem titled *"The Orchard"*. Four maidens stand with scrolls of text in an orchard with a flowery fore-ground.

The oriental influence was strong in gardening design, it reached Europe in weaves; the collector of chinoiserie saw the Chinese gardens on screens, lacquer-ware. Travelers brought the ideas to Europe with them. The Chinese garden is surrounded by a wall with a gate. Covered galleries connect the garden pavillions, along steps, walks trees, rocks are planted asymetrically. An old Chinese imperial lacker box has a garden scene on top; a group of Chinese connoisseurs are looking at a scroll on a terrace. Two young attendants stand by with more scrolls. Behind a few steps lead into a garden pavillion with laticed walls and double curving roof shaded by old trees of distorted trunks. Rocks are like distant mountains, clouds descend over them. The Chinese garden is very different from the European garden; it is not possible to understand it without long training in Chinese philosophy, Tao and Yen-Yang. Chinese gardens are built to symbolize ideas. Taoism and Confucianism both searched for an inner reality in the blend of opposites, the union of extremes, to find completeness.

The polarities of mountain and water must be there to meditate upon, in solitude. Strange hollow rocks stand for mountains. Rocks are often layered in a horizontal or vertical way. Sometimes artists improve on nature and help to shape the rocks to make them right. A rock is part of Tao, the created world. The Chinese garden is an understatement; its beauty may be appreciated only by the cultivated mind. Special literary knowledge is needed to understand its symbols. An outsider, walking through it can not grasp its beauty. Changing seasons, weather, the hour of the day brings out different aspects of the garden other times or conditions create other moods — certain points of the garden are specially suited for one or the other mood.

Hangchow was the capital of the Kingdom We-Yueh (907-78) during the period of the Five Dynasties (907-60). The rulers had a luxurious, refined taste and were great patrons of the arts. During the 10th century it became customary to mount paintings on brocade. Monochrome paintings reflected the ideas of purity and simplification of Buddhists and Zen. Artists specialized in flower-and-bird paintings and the first pure landscape paintings were born. During the Sung Dynasty (960-1279) Northern China was attacked by invaders. A member of the Imperial Family finds refuge in Hangchow and once more Hanchow is the capital, protected by the Yangtze River. The Academy of paintings is founded in Hangchow. The Northern Sung school of painting can be recognized by mountains upon mountains, immense, limitless nature. In the Southern Sung a cloudy mist expresses the infinite — one lonely boat, a mountain above the river banks. Up to this time, Hangchow was the Venice of China; full of lakes, rivers, bridges, pagodas, small waterfalls, islets with luscious vegetation of the South.

These elements; the gardens, bridges, pagodas became a "pattern" by the Chinese ability to transform elements of nature, clouds, trees, rocks, mountains, into ornament. Such little landscape-ornament, when repeated as textile pattern is used for silken brocades produced in Hangchow.

The Ming Emperors moved to Peking from Nangking in 1420 and the Imperial City of Peking was built. Walls within walls surround the Imperial compounds, the "Forbidden City" — closed for centuries to foreigners. Gates, palaces, audience halls are followed by the private quarters of the Emperor and the concubines, enclosing small gardens and looking out at the Imperial Gardens with bridges, sculptored ballustrades, pavillions. Windows and openings of the walls were of different size and shape; according to Chinese ideas the same view seen through different shapes will always be something new.

Ancient myth, astrology, religious and aesthetic rules are followed in China while placing a burial place, house, gate, or garden. The four directions correspond to the four seasons, and their symbols. The main road of the "Forbidden City" is laid out in a symmetrical North-South direction intercepted by an artificial river, a marble lined canal with five marble bridges, a deliberate construction made to resemble a piece of virgin nature.

We may compare the "Imperial Garden" of the "Forbidden City" with the Garden of Versailles; both are at the highest level of perfection, refinement, a peak in civilization and artistic achievement; their outlook is totally different. In traditional Western philosophy man is the center of the universe. To the Chinese man is part of nature together with rocks, plants, animals — to understand nature is to understand the universe. Nature symbols are basic in Chinese art.

The Chinese used to call their country the "Flowery Land" — the beauty and perfume of flowers are a part of its life. Weddings, funerals, festivities could not be held without them. Each month is symbolized by a flower — and an expert can tell of a few flowers whether the painter wanted to represent a windy afternoon in spring or an early morning at the height of summer. The transformation of the flower from closed bud to blossom indicates the beginning or end

Byzantine Empress Theodora Domjan

of the month. The hour of the day also changes the condition of the flower; during noon hours flowers are wide open, petals are loose in the palor of heat — buds close, blossoms are fresh and moist during morning hours — a drop of dew on a petal that seem incidental has a meaning for the Chinese. To be able to detect these fine nuances 3000 years of observation and contemplation of flowers are needed.

Silk, to the Chinese, who originated it, to the Byzantines who used it lavishly and before to the Romans, silk was more precious than gold. The Greeks called the Chinese *seres,* silk people. Aristotle wrote about silk worms as "a great worm that has horns". Silk thread was used for tapestries mixed with wool from the fifteenth century. Chinese tapestry is woven of silk thread.

Silk was a guarded secret of the Chinese for more than two thousand years, since the wife of the emperor Huang-Ti started sericulture, even invented the loom, the delicate work of silk production was in the hands of women in China. Women collected the silk moth in baskets, where they laid their eggs; then gathered and chopped the leaves of mulberry trees to feed the worms' enormous appetite as these emerged from the larvae stage. The mature worms' glands secrete the silk, they wrap themselves into cocoons, these are then boiled and the silk unreeled. The silk worm is so delicate and sensitive the quality of silk is influenced by noise, even smells.

The legendary silk route, a strech of 4,000 miles from China to Samarkand, Baghdad. Palmyra, Antioch and then by boat to Rome was the longest and most adventuresome trade route in history. From the heart of China, (Chang-an, sera Metropolis) the silk was carried by thousands of camels through deserts and mountains, changed hands by merchants of many different cultures and many tongues.

In 552 two monks, Byzantine ambassadors, sent to China, smuggled the eggs of the silk worm out to Constantinople in the hollow of their cane. Those eggs started silk production on the European continent.

Domjan during his trip to China went to Hangchow to the weaving workshops that produce woven silk cloth with the same method as in ancient times. The most complicated pattern weave is made by hand. On a large wooden loom, a complicated construction, a 36 in. wide fabric was woven with the traditional ancient patterns in silk. Two old weavers work at the loom. One sits below, looks up, his eye blinking; the top of the loom is in darkness, one can hardly see, but up there sits another old man, directing, selecting, lifting the warp threads according to the pattern he has in his memory. The silent signals, the cooperation of the two old men — their knowledge of intricate patterns in minute detail — is remarkable. Chinese workmen in the old times had three days off for the New Year, otherwise were expected to work every day of the year without vacation except for the day the man was married or had to bury a close family member.

Young men cannot be trained for this ancient silk weaving any more and when the old men die out the method has to be changed. The patterns on the looms were the landscape of Hangchow; the Venice of China with many lakes, bridges and islands; pagodas and gardens. Ancient symbols were also woven; the dragons, birds, sacred discs. In ancient times the silks were used for robes of the Emperor, the mandarins, then costumes for the traditional Peking Opera. The smallest piece is of great value; it is often framed and hung as a picture and is used for binding for sumptuous volumes of Chi-pai-shi color woodcut prints.

Chinese tapestries, called "K'o - ssu" or "woven picture" are always silk, they have nature, flowers and birds as subject matter. The small, 3 x 4 feet tapestries are sewn together to make a large hanging. Panels were used to decorate the Imperial Palace and were woven in the Imperial Workshops. Two dimensional in feeling, Chinese design fits the style of tapestry weaving perfectly; the backgrounds are solid blue, dark plum, wine, peach color with empty spaces left around, blossoms scattered according to the Chinese composition where the center of the composition is outside the frame and not in the picture — the composition of the infinite . . . fine weaving, shimmering, silk, delicate details, mastery of observation of branches, feathers, decorative handling of leaves, give the Chinese tapestry its unique charm. There is a clear definite contour line, curving lines are used for the veins of leaves, petals; broken, straight lines for the strange characteristic rocks, one big hollow rock, or a group of rocks. There are the rocks, the water, in undulating woven shapes and lines in a decorative way; flowering branches and bamboo leaves — chrysanthemums, magnolias, peonies, plum blossoms, pine trees — and the animals — ducking pheasants, from dragons to butterflies and the Fen-Huang. This is the repertoire of Chinese tapestries.

Imaginary Animals

Kachina Doll Šomjen

Celestial Unicorn

Birds as Symbols. In medieval art, birds were the symbol of the soul, and the symbol of freedom. The dove played an important part as representation of the Holy Spirit. In hymns, in love songs, in embroidery, and in tapestry, birds were a favorite subject of medieval artists.

Drawings of birds abound in the "exempla"; these in turn were used in between vines, grapes, flowers and ornaments on pages of manuscripts and were woven into "milles fleurs" tapestries. Both domesticated and wild birds are found in the "exempla", but not all birds were drawn from life. Some came from a collection of ornaments and stylized elements. A siren with a woman's head and bird's body from ancient Greece easily made its way into borders of illuminated manuscripts. Among birds drawn from life are those seen around the house — sparrows, chickens, cocks, storks, ducks, crows, magpies; with hares, rabbits, squirrels, and insects shown promenading in the grass or picking at berries in the "milles fleurs" tapestries.

In the *Hunt of the Unicorn* Tapestry at the Cloisters, there are at least a dozen different species, thirty birds in all, woven after naturalistic drawings. Not only are the birds recognizable, but they are in their natural habitat; aquatic birds in the pond or river, with marshy plants; perching birds on branches of the trees. They are used to enrich the design and perhaps they also have a symbolical significance. At the tapestry with the fountain (second in the series) there are two pairs of birds who came to drink; a pair of pheasants, and a pair of gold finches. The goldfinch was a symbol of the Crucifixion and the Passion because goldfinches like to eat thistles. Near the fountain in a bush is a nightingale. It was believed nightingales bring luck to weddings, and since the tapestry series was made for a wedding present, the nightingale's presence is appropriate. Birds without symbolical explanation on this tapestry are a woodcock and a bittern, with rust-colored feathers; on the next tapestry, a pair of partridges, a mallard and a domestic white duck. In the fourth tapestry with more mallards and woodcocks there is a purple heron. The legendary phonex bird may have been a purple heron, "phoeniceus" means purple. However, the bird in this tapestry is more like a common grey heron, frequently found in Flemish marshes, or near river banks. On the last (sixth) tapestry there are two swans, harbingers of good fortune, and doves which are also associated with mythological and religious meaning.

In ancient China the crow was the symbol of the sun, personified in animal form, while the moon appeared as the hare or toad. The phoenix with peacocks and mandarin ducks appeared on Tang mirrors in connection with marriage ceremonies. Pairs of birds in Chinese paper-cuttings — a folk art flourishing even today — are pasted on the windows of the newly wed to bring luck.

Man and Bird: Birds were symbols, and once in a while man became a bird himself. Feathers were the earliest decoration of the males in primitive tribes. Medicine men used bird masks and feathers, like the American Indians in their feathered outfit they identified themselves with birds during ceremonial dances and magic. Man's eternal wish to fly must have something to do with it and the decorative beauty of the feathers. Australian aborigines pasted feathers according to a design on their bodies and in Peru feathers were woven into tapestry. One of the great acts of feather fashion was the 19th century; at this time it was the ladies who wore the plumed hats, ostrich boas, fans.

But the real bird-man was the shaman. The shaman was a messenger, who could fly and run, and make a trip to other worlds; his costume was an animal disguise and the metamorphosis of man. The shaman had super-human talents; he could travel to heaven and to the bottom of the sea; to do so, he identified himself with birds. Shaman ceremonial robes had wings, were built on a bird-shaped, copper framed skeleton, rather like a crinoline hoop skirt, covered with a fabric; on this, small metal discs were sewn with magic symbols of earth and heaven, eagles, bears; also bells, shells, feathers and mirrors. The shaman's shaft had a horse's head.

Shamanism did not die out. The shaman still performs his healing and super-natural rituals against official rule in Mongolia. A large fire is lit in the open with spicy, perfumed sticks and the shaman, dressed in skins, dances to the beat of a drum. There has to be a sacrifice and the participants, one by one, throw a precious possession into the flames.

The Domjan bird tapestries. In these tapestries pigeons, peacocks, eagles, and phoenix birds transcend their subject. In Domjan's color woodcut works the birds came early. While flowers were made after nature the birds came always from imagination. First a series of small pigeons, then they got more complex and grew in the process. A little woodcut of a pigeon and dragon was a tremendous success through a series of one-man exhibitions in China. In the United States, an exhibition called The *Peacock Festival* was shown from coast to coast and traveled for years. In the tapestries, the birds culminated in the *Fire Peacock,* ten feet long. This is certainly a peak in Domjan's work.

Topaz is a fabulous bird, with oval-shaped tail plumage and a straight long neck. An ornithologist may speculate whether it is perhaps an ostrich or a mutation between a giraffe and a turkey. Whatever it is, Topaz is a dominating presence with a sense of humor in its astonished blue eye. Enveloped in stylized foliage and tulip motifs, this truly fantastic bird comes from a land where there is no border between imagination and reality. Floral elements appear in lighter and darker turquoise on the bright amber bird and in muted shades of olive green and blue-green on the dark, almost black background. The colors and dimensions of this background foliage are very much like the "verdure" while the zig-zag, triangle and scroll motifs are Domjan's own language. This "verdure" background is a striking combination of the new and the traditional; it was woven with gusto in Aubusson.

The crest of the bird is of seven stylized peacock feathers, two of them stand out of the oval shape; a particle of the large feather blown away — stands alone, a bright cadmium drop in the dark "verdure"; it causes the eye to return to the center of the composition. The stem of the foliage in the bird's beak also circles the large central shape. A faint carmine brushes the top of the bird's plumage like an Alpen Gluhn. The interplay of turquoise-yellow-green-rose gives a luminous effect as if the bird had been lit from behind.

This bird is larger than life-size, one has the feeling it is a disguise; the shaman (priest, wizard and medicine man of Ural-Altic nomadic tribes during the migration of the peoples) is behind it, and looking out of one of the dots — it can unfold its wings, fly away — be here and at another place at the same time — be any place at will. The shaman's bird-shape ceremonial gown must have been heavy; and his movements slow and majestic, yet he could create the illusion of speed and lightness. The tiny mirrored discs reflect the sun, the bells tinkle, the feathers are blown in the wind.

Peacock of Carnations. A monumental golden bird fills the space. Its tail is fan-shaped, in seven parts, and its body and the background are strewn with stylized carnation motifs. Carnations grow on straight stems on the tail, and each is a substitute for the round eye of a peacock feather, also on the three-fold headdress. On the body of the well-fed bird, in the background, carnations are curving, intertwined. The golden bird is cadmium light, with carmine edges; the cobalt blue carnations on it and the ultramarine flowers on a darker background of blue-grey colors are in perfect balance with the complementary colors of yellow and blue. In this tapestry a simple principle is achieved with the most sophisticated diversity of tonal variations. There is a feeling of spaciousness in spite of the fact that the peacock shape goes from edge to edge and there is a feeling of depth in spite of the two-dimensional ornamentation. The animal is in a blue jungle of blooming carnations, which are both outside and also in its body. A fabulous creature from the land of dreams.

Moonshine Peacock. A large static shape, an enigmatic resplendent idol is caught in stylized plant and animal elements. Repeated butterfly motifs, like a flower, attached to the trellis by a stem, decorate the body and "fly" around it. The bird body and fan-shaped tail are in two colors; light yellow in three shades, fading into four shades of cream-ivory; and azure blue in five shades. The background of several shades of umber-into-grey are made of distinctive battage of close hues. There are clear, decorative contour lines. Small triangles, like so many holes, have an opposite color shining through, like a net, yellow on blue, and the dark umber background through ivory. *Moonshine Peacock* transcends time and style and is equally at home in period as well as in up-to-date rooms.

Le Coque Rose is a graceful, joyous bird in delicate cold rose-grey colors on a dark green background. Inside and around the cock are stylized plant and bird motifs. There are some intimate details; the dark purple little house, the milk jug, also a pair of mating birds in dark carmine red with purple and cadmium. The Domjanian little details dots, squares, ornaments are playful, as though drawn by a child; squares, checker boards, criss-cross netting, diamonds, unlike in any other tapestries. The color composition is also unique; if one looks as the "echantillon" of colors (this is the long string of wool the weavers put together after the dyeing has been done and send it to the artist on approval if he is not around to see) the three shades of lilac, two greys, three rose colors, strong reds and purples and the dove-colored pale shades are clashing and disharmonious separate, but in the right doses and at the right spot and relationship with the colors around on the tapestry, they sing together in unusual beauty and harmony.

A fish swims in the water, a bird flies in the clouds. A drawing is like a dance, it is movement; but the dance leaves a trace on the paper. The weavy, swinging lines of *LeCoque Rose* can be followed in swinging gestures and beautiful posture. This cock seems like choreography.

Turtledove. This is a unique Domjan design of a dove with his three wings extended and small birds sheltered under them, like one of the "milles oiseaux" tapestries. The bird in pigeon gray and blue with green looks very soft; the

63

small birds are in slightly contrasting light green malachite on a rose background. There is the solid, simple shape of the large bird's body and there are the small birds. This creates a strong silhouette effect. The birds are strangely arranged; some stand, others are in a horizontal position in different sizes, shapes and ornamentation of broken-up forms. An almost solid faded pink background recalls the rose meadow of the *Lady of the Unicorn.*

The Blue Cock. A big wave engulfs it and the cock comes out triumphant, proud, undaunted. Its red crest is shining, a shrill voice against the pounding waves. Small fish swim in its tail, forming contrasting lines, rhythmic wavy ones and straight, correct parallels, interlaced in curving; graceful calligraphy. The color, the movement brings back the sea; but not the sea alone; there are star motifs, distant and reflected in the sea, a variation of the star motifs. The *Blue Cock* has a complex ornamentation and a brightness within a limited palette. There are eight blues from turquoise to ultramarine on the body and three reds for the crest.

Eagle. This is a strong, dominating upright bird. While usually there is a colorful bird on a dark neutral background; here is the opposite: a dark blue-black eagle stands on a bright red background, but still the bird stands out in the foreground. The moon cuts into its wing and below there is another moon — a moon eclipsed with stars in several shades of bright yellow, indicating the bird's cosmic rule. These sparkling star-studded designs, with moons and galaxies characterize Domjan's latest work.

His Majesty. A royal personality with a graceful head on a swan's neck and a crown of peacock feathers. This is a purple-blue bird with open wings, air born — Is it a peacock or a phoenix or even a Fen-Huang? Whatever it is, he is a majestic visitor. There is a carmine shape around his breast, like fire or blood. Is this a swam song — are we the last to see this delicate silhouette against the umber background?

Midnight Phoenix evokes the mood of the lake on a cool summer night, the silence, the soft splash of water and the silver skies reflected in it. Simple and harmonious in form but complex in details, its star motifs fill the phoenix bird and the dark background around. *Midnight Phoenix* is like a cobalt blue glazed Ming vase, with many shades of blues, deep and transparent. An art work is not just a beautiful thing to decorate a wall with, a valuable *objet d'art,* it is rather a need because of the spirit it emanates.

The Phoenix. In the church of St. Appollinare in Classe, Ravenna, there are mosaics at the St. Agnes altar built in 700 A.D. Eight days after her martyrdom and death, St. Agnes appeared to her parents, splendidly dressed in bridal garments, and on her magnificent robe there was the phoenix bird, symbol of resurrection and transfiguration.

The phoenix, a fabulous bird, sacred to the sun, like the unicorn, is one of those creatures that has fascinated people both in ancient and in modern times. To doubt its existence was impossible because it had been so much involved in Classical, Rabbinical, Christian and Islamic legends. One phoenix feather was sent in 1599 to Pope Clement VIII and as late as 1840 a highly esteemed scholar at Oxford still seriously believed in the existence of the bird. It did and does exist, created by imagination and kept alive by fantasy.

The phoenix name comes from Greek, meaning "bright colored". The word

also may mean "palm tree." A large number of classical and post-classical authors from the 5th century B.C. among them Herodotus, to the Middle Ages give the following description of the phoenix; "The phoenix was a large bird of very gorgeous plumage and sweet voice. It was always male, the only one of its kind and lived very long." Various authors give the period from 500 years up to 12,954. Tacitus gives 1,461 years, which is the Egyptian Sothis period in which the year of 365 days circled in succession through all the seasons. (The tropical year, determined as it was in Egypt by the heliacal rising of Sirius, Sothis, was almost exactly the Julian year of $364\frac{1}{4}$ days, differing from the true solar year, which was 11 minutes less. The Sothis period was thus 1,461 years.)

At the expiration of his time the phoenix made itself a nest of twigs of spice trees, on which it died, by setting the nest on fire and burning itself alive. From its body, or its ashes, or the nest, which it had fertilized, came another phoenix. The young bird took the ashes of the body of his father, covered in spices, and the nest and flew with them to Heliopolis, in Egypt, where the young bird deposited them on the altar of the Sun. There are many variations in details.

The phoenix legend or saga is as old as the human race. A fabled bird, under various names was familiar to Egyptians, Greeks, Romans. St. Ambrose and St. Bede show how the great bird captured the imagination of the early church. On a fifth century mosaic from Antioch, in the Louvre, the phoenix was the symbol of Righteousness who arose with healing in his wings. The phoenix, a sacred bird, is standing alone on top of a mountain. Around his head a halo or sun radiates in all directions.

The Chinese phoenix, the Fen-Huang has a cord in his mouth symbolizing the marriage bond. It is seen on ceremonial wedding mirrors. The Fen-Huang is the subject of silken tapestries; he flies in the clouds, is seen among other birds as their king; his neck is like the flamingo's; his tail plumage is long and flying after him, delicate ribbons; his plumage is complex, often his out-stretched wings cover the picture space.

Phoenix on his Nest, tapestry by Domjan. The large boat-shaped horizontal orange body of the phoenix with seven feathers nests on what one imagines to be perfumed twigs. There is an orange radiation of the sun, a radiation of flames. Stars sparkle, illuminate the night in a golden glow. The composition is of a dark and bright contrast, positive and negative areas of hill and mountain divided by a clear curving line.

Centauress

The Unicorn. On a green island against a rose background sown with flowers in the "milles fleurs" style, a graceful lady plays with a unicorn. *The Lady with the Unicorn* series, now in the Cluncy Museum, reflects the elegant, romantic worldliness of the period. The series is also called the six allegories of the Virgin, or the senses.

The unicorn, a fabulous beast, with a long, single twisted horn in the middle of his head, usually has the head of a deer, the beard of a goat, the body of a horse, the tail of a lion, although this combination changes. Sometimes it is the head of a horse, hind legs of an antelope, etc. Aristotle mentioned one-horned animals, the oryx, a kind of antelope, and the wild Indian ass. Pliny also writes about it. Travelers had their strange stories of far away animals, also it could have been an error in translating the arabic word for rhinoceros. In any case the unicorn kept its place in arts, in imagination, even in science for a long time.

Miraculous talents were attributed to the unicorn. The medieval concept was that it was very swift, impossible to catch. Its wildness and fierceness could be tamed only by a virgin. All the dogs and hunters could not catch the unicorn, but the virgin goes into the forest and puts a band around its neck and leads the unicorn into town. The virgin, possibly a nude, often rode into town on the unicorn. The unicorn lived alone in a secluded place.

When the unicorn put his head into the water, the water became clear. The beasts of the forest waited at the fountain for the unicorn to cleanse their drinking water.

Powdered unicorn horn was more expensive than gold, was supposed to cure madness, the plague, and was given to royal children like vitamins.

The unicorn horn (narwal tusk) of the Cluny Museum comes from the Abbey of St. Denis. Here it was described in the inventories of the Sacred Treasury and Holy Relics as one of the most precious objects sent to the Emperor Charlemagne by Harun al Raschid, about 807 A.D.

Unicorn horn was used as decoration on drinking cups to combat poisoning. During the reign of Charles II, King of England (1630-1685), a cup supposedly made of the horn was given to the Royal Society for experimentation; after that, it was announced that the horn did not have the anti-poison properties attributed to it. However, as late as 1789 the horn was still in use in the French court for testing royal food for poison. The royal throne of the Danish was said to be made of unicorn horn. The unicorn horn in the Falconeri coat of arms has the motto: "Semper purus." In heraldry the unicorn is used as a device; in England it still is the left-hand supporter of royal arms.

The Lady and the Unicorn series was woven in 1495, probably in Aubusson although we do not know for certain. Perhaps it was woven by an itinerant weaver who, with his bobbins, comb and a letter of recommendation, went from castle to castle for temporary work. However, it is quite possible the tapestry series was ordered for Claude Le Viste, as the Le Viste coat of arms dominates each tapestry of the series. The tapestries were in Boussac castle when George Sand wrote about them in 1847.

eagle

The Lady is a legendary personage living on a fairy tale ring in a rose world. So strong is the spell of this imaginary world that we are not astonished in the least at seeing flowers bloom on a rose meadow.

On the tapestry allegedly representing sight, the Lady sits in the center dressed in a richly brocaded gown. She is holding a mirror; the unicorn — a large, white, proud animal — his front paws in the Lady's lap, is looking at his image in the mirror. Here is the friendly lion holding a banner with three half-moons, insignia on a ribbon, on the banner. There are other animals on the tapestries. Always present are the monkey, the lamb, the weasel, the fox, a little bear, the crane, the hare and dogs. The flowers are lilies, iris, daisies, clover, violets, pansies, herbs, grasses, wild flowers. There are trees too, — oaks, mulberry and apple.

The theme of the last tapestry is homage to the Lady's beauty and virtues. A silk pavilion is the center with the inscription: "Mon seul desir." The Lady in a turban-like head-dress is choosing jewelry from a coffer presented by the girl attendant; standing on guard, the lion and the unicorn are holding the silk curtain on both sides.

Who is this serene, reserved, delicate and enigmatic beauty? Is she Lady Blanchefort whose sad unfulfilled love for Zazim, son of Mohammed II and prisoner in the tower of Boussac, was sung about by the troubadours? Or was the heroine an imaginative personage? We may never know.

The Unicorn Tapestries in The Cloisters. Also in the "milles fleures" style is the admirable series of tapestries in the Cloisters, representing the hunt of the unicorn. These consist of six tapestries and fragments of a seventh. Borders and probably the upper part of each are missing. After mutilation, bad care and centuries of use, the tapestries shine in indestructible beauty. During the French Revolution they covered potato bags to prevent them from freezing. The Edict of 1793 ordered that tapestries with the royal insignia be destroyed. We are grateful to the unknown Frenchmen of Verteuil who, perhaps facing danger of punishment, defied the order, and cut away the insignia, thus saving the tapestries.

The tapestries were woven for Queen Anne of Brittany, widow of King Charles VIII of France, as a marriage present from her second husband, King Louis XII (1462-1515). The marriage took place in Paris, January 8, 1499. The original series consisted of five pieces, the first and seventh are of later origin, added by King Francis I who was married to Queen Anne's daughter and heir in 1514. Ferdinand de la Rochefoucauld received the tapestries from his godson, the King, and they stayed in the family in Verteuil, where they are included in an inventory of 1728, until purchased by John D. Rockefeller and given to the Cloisters collection.

Queen Anne's colors as well as King Louis' colors are used. The letters A and E (first and last letters of Anne, the E in reverse) tied together with a cordeliere in red white colors, are used as ornament in large size and repeated five times in the milles-fleures background of each tapestry. The subject matter is the hunt of the unicorn, and also the Incarnation of Christ. The monograms are included with an allusion to courtly love ending in consummated marriage. In the Middle Ages way of thinking, symboloic hidden meanings were enjoyed and contemplated upon.

The Unicorn Tapestry by Domjan. A fine well-balanced shape, square body, long neck, its head turned back, bearded, the long unicorn horn in a diagonal, the unicorn is looking straight into your eye and follows you with a gentle look of his carmine red eye. Its front legs are lifted — about to leap — the tails, he has two of them with a mop at the end of each, curved across a large half moon. The unicorn looks around: Did I lose my tails in the moon by any chance? The animal's body is ocher and reddish in many fine shades. The moon is pale silver, aquamarine. The background is dark, bluish filled with stars, planets, milky ways, polar systems; a whole universe. There is not a half inch on this tapestry that is not broken into so many fine soft shades, small forms, short battage of close colors, fine, hardly visable transitions, hachures of finest hues. Stars, moons in the Domjan style, broken up into endless smaller stars within stars, moving, spiraling, circling, radiating. The eye is captive, follows the galaxies around the red eye.

A truly unique Unicorn.

Bestiaries. Imaginary animals, hybrid forms, monsters populate the arts since the cave paintings. A variety of creatures were invented by fantasy and believed to be living. Their shape was composed of human and animal forms like the sphinx and the centaur; or purely animal combination like the unicorn; there were even animal and plant forms combined like in the grotesque, or initials of manuscripts; or the aquamanile, copper or brass water jugs in shape of fantastic creatures.

Seafarers had their legends of sea monsters, eating boats, like the Phoenicians to frighten away commercial rivals. Animal-headed masked priests of Crete spread the idea of metamorphosis; the Mediterranean preserved the Minotaur cult in the bull fights. Imagination was regulated in classical Greece by Sirens, half bird half women creatures, who lured Odysseus with chants and magic song.

In China the dragon was one of the four fabulous beasts associated with the seasons, and creation, with the tortoise, the phoenix bird, and the unicorn, or tiger.

Oriental motifs spread through Sassanian art and found their way into carolingian manuscripts and stone carvings of the Romanesque and Gothic art. Officially ancient Rome was against the supernatural, "monsters that are not, nor could be, nor ever were" (Vitruvious). Similar were the voices during the Middle Ages against the fantastic animals in the cathedrals — "what are they doing there, diverting the attention of the believers" . . . St. Bernard of Claireveaux writes about those ridiculous monsters, marvelous and deformed creatures, the centaur, four-footed beasts with serpent tail, fish with beast head — if man is not ashamed of his follies he should at least economize — had these voices been listened to we would be all that poorer. During the early Middle Ages fantastic animals were used as a decoration, 13th century scholastic thinking attributed meaning to each creature, explained, analyzed. By the end of the Middle Ages fantastic animals were again used for their artistic decorative quality alone.

Medieval bestiaries are a collection of myth, travelers' tales, folklore, moralistic teachings, and previous accumulated misunderstandings mistranslations. These descriptions were in turn illustrated, and more imaginative creatures were

71

added to the zoo of fantasy as artists tried to follow the literary descriptions of animals. Animals were attributed vice and virtues and analogy was made between one or the other animal and Christ.

Exempla. The collection of drawings used by craftsmen was called "exempla" and served as a means of transmitting traditional forms. The artist-craftsman who made the drawings concentrated on the illustrative elements and not on the style of his subject, and so used his own interpretation. When we are looking through the pages of the exempla, we cannot tell if a motif came from an architectural decoration of a cathedral, or a piece of silk, or if the original was in miniature or life sized, if it was stone or gold. The exempla served as a source for miniature painters, for sculptors, and goldsmiths, and tapestry weavers.

The earliest sheet of exempla now in the Vatican Museum is of 10th century Rheimish origin and is inserted in a Carolingian manuscript. The Einsiedeln Codex of the 11th - 12th century is the earliest known complete notebook for illuminators and craftsmen. Few of the drawings were taken from life. These artists were trained to draw only stone lions, not ones that move.

The "exempla" was one of the means by which the unity of the International Gothic style was achieved — motifs carried from country to country, with migrating craftsmen, a certain eagle (perhaps Sassanian) from a portal of a cathedral to the capital of another in another country, to a silver reliquary, to the decoration of an initial in an illuminated manuscript to the cartoon and woven into a tapestry . . .

Devilish Deer Tapestry by Domjan. A graceful animal but with horse hoofs, the head of a goat, a very large slanted eye, ruby, (seen in profile) its tail ending in an arrow shape, has antlers bursting into flower forms; the light filters through — flowers, or stars — jade green and light amber yellow; smaller units fall off and the wonderous stag steps over them, he walks in the skies with blooming head gear. The animal itself has that indefiniable Domjanian cloud colors — grey-blue-turquoise-brown-pale shades. The background is almost solid dark blue in four shades.

Gentel Dragon

The Magic Carpet

Grotesque

The Magic Carpet. Tapestries, their images vivid in color, warm cold stone castle walls. Tapestries, telling stories, brighten Romanesque churches on holy days; tapestries, blowing in the wind, create a festive accompaniment for processions, tournaments; tapestries celebrate royalty, visiting dignitaries. Tapestries, used as curtains on alcoved beds, bring hope and cheer to their beholders in medieval burgher homes. Areas enclosed by tapestries for intimacy become delightful refuges within vast castle halls. Tapestries used as bed dividers bring solace to the poor suffering in hospital wards. Tapestries are flaunted proudly as majestic throne baldachins. Tapestries, rolled up or folded, packed in carriage trunks, on mules or camels, travel long distances. Medieval princes carry their tapestries, their most personal and precious belongings, with them from castle to castle, from battlefield to battlefield. As princely gifts, purchases, or booty, tapestries emigrate to foreign lands, disperse.

Inventories. Inventories made during the Middle Ages are valuable even though inadequate documents. In most cases there is no exact description, no size given of a certain object. They do not say of what it is made, much less when or who made it. Even though a name is recorded that may indicate the patron who commissioned it or who supervised the work, the question remains — who designed it and who actually made the tapestry?

Inventories were taken at the Chateau Nancy, residence of the Ducs de Lorraine. In 1530, and later, the lists included treasures, arms and armour, furniture and tapestries of religious, historical and myth-subjects: *History of Moses*, 9 pieces, made in Lorraine; *Life of Abraham*, 10 pieces, made in Brusselles with the coat of arms of Cardinal Charles de Lorraine; *Life of Alexander*, 11 pieces from Les Gobelins; *Scipio Africano; Romulus and Remus; Hercules.*

According to a list of 1540: In the room of Monsieur seven pieces, verdures, 12 pieces, including four we have cut (!) to serve where needed; 23 pieces of verdures, some big some small, including two which are at Madame Marie's; 8 pieces of verdures with beasts, 8 others where there are no beasts; 8 small old tapestries of no great value; 6 couvertures de mulets of no great value, etc.

Here we have proof that tapestries were cut to fit a place. The old ones that were not valued may have been the most valuable. We do not know the subject matter of those that have no beasts. All of these are lost or cannot be identified.

The court of Nancy's tapestry collection was under the charge of de Jehan, a tapestry worker, who was probably trained not only to hang the tapestries but also to restore, repair and storage them. This must have been an imposing collection with several series and several workshops represented.

In 1499 Philip the Good of Burgundy ordered the *History of Gideon* tapestries, an eight piece series, from Tournai after cartoons of Baudin de Bailleul who was the most famous cartoon painter of his time. The *Gideon* tapestries decorated the meeting hall of the Order of the Golden Fleece and were used at a number of celebrations: Philip the Good had them at his residence at the reception for Louis XI in Paris in 1461. They were used at the marriage ceremony of Eleanora, daughter of Philip the Fair and Johanne of Castilia in the Saint Judul Church in Brusselles in 1498; at the abdication of the Emperor Charles V in the Castle of Brusselles in 1555; at the marriage of Alexander of Parma with the Infanta of Portugal in Brusselles in 1565; during holidays in the

royal Chapel in Brusselles until the end of the rule of the House of Austria in 1794. Then the Gideon tapestries disappeared.

The Duke, Philip the Good, had a large collection of tapestries. Mr. Jean Aubry, "grade de la tapisserie du Duc" — was curator, caretaker; he ruled over six guards and twelve servants. In the Hotel Amboissenelle in Arras, there were two large rooms, whose armoires were locked with keys; the tapestries were kept in these rooms. The curator-guard slept in these rooms. Another room had a stone vault constructed in 1444 to save the tapestries in case of war or fire. In 1469 the guard, Garnelot Poucelet, was ordered by Charles the Bold to make a new inventory of all the tapestries.

Historic-Classical Subjects In Tapestries. The Trojan War, antique history, Alexander, mythological themes are frequently used in tapestry; Hector and Darius are well-known heroes on the walls of castles. Medieval peoples were looking for their ancestors among mythological characters. The Franks claimed to be the descendants of the Trojan Frankus and the English dynasty of Kings of the Trojan hero Brutus.

Christian Church fathers placed antique heroes in the same category as patriarchs. See the *Nine Heroes Tapestry* series in the Cloisters.

In astronomy, pagan gods ruled as inhabitants of celestial bodies: Jupiter, Mars, Venus. This was a naive knowledge of antiquity, derived from legends spread by troubadours and jugglers.

The jugglers sang about historical events from memory. The jugglers changed historical dates; mixed history with marvelous fairy tales; through fusion of several personages great deeds of several generations were attributed to one hero.

The robust stone walls of the fortress-castle enclosed a large hall where all gathered to hear the jongleur; the prince, the chevaliers, the valets, a traveling monk, merchant, beggar. The stone walls were hung with war trophies, heraldic symbols, arms and armour that did not get dusty, they were always ready for a new war or tournament. And the tapestries were on the wall telling of stories like the ones sung by the troubadour.

One of the most popular novels of the Middle Ages was *Alexander,* a poem of 20,000 lines. In it historical facts were mixed with Oriental tales, the war of Alexander against Darius was described as the war of the times. The most fantastic is the description of India, land of marvels, imaginative flora and fauna. There are also episodes of impossible, Julius Verne-like forerunners. Alexander is flying on a griffin and he descends to the bottom of the sea in a glass bell. Alexander was the most fashionable, perfect knight, a great celebrity of the 12th century, and so he became a personage of tapestries and a favorite with jugglers. There was no historical sense in the verses, so it followed, naturally, that on the tapestries Hector, Alexander, and Darius were dressed in medieval attire. Greek warriors were shown as armoured soldiers of the Middle Ages and Troy was depicted as a medieval fortified castle. In 1459 Pasquier Grenier, Flemish weaver, made a series of Alexander tapestries now in the Palazzo Doria, in Rome, with kings and queens in hermelin cloaks, with armoured soldiers and horses. Another favorite story was the war of Troy. A long rhymed poem also written in the 12th century about the seige of Troy was enriched by countless episodes of courtly love, gallant wars, and great heroes.

The *"Destruction of Troy,"* the most popular woodcut-illustrated book, published in nine editions, between 1484-1526 continued to perpetuate false images of the classical world.

Caesar crosses the Rubicon on the tapestry in the series of *Caesar's war against Pompeius.* The series of four pieces of large size (4.30m x 7.56m.). in silk, shows many figures, woven in Tournai (1465-1470), for the throne room of Charles the Weak; once in the Lausanne Cathedral, it is now in Berne. I have crossed the Rubicon several times and rarely noticed it leaving the outskirts of Ravenna. On the tapestry there is a medieval walled-in city in the background — I wonder if it could be Ravenna? Medieval soldiers, heavily armed, mounted on armored horses are shown in deadly battle. Some are falling off their horses, with head or limbs cut off; missing body parts are scattered in the "milles fleurs". It is a puzzle in crowded confusion.

Popular theme of tapestries from the thirteenth through the nineteenth century was the allegory, a personification of ideas. If a shapely maiden is Virtue or Fortuna, it doesn't make a great deal of difference to us, but in the scholastic way of thinking peculiar to the Middle Ages, it was very important. The novel that enjoyed unsurpassed popularity from 1230 was *Le Roman de la Rose.* Maymorning walks in the country, arrives at a walled-in garden; here are Pleasure, Love, Beauty. The rose bush falls in love. Danger, Fright, Jealousy, Chastity, Honor, Respect play roles in the novel. There were manuscripts in several versions of this most popular novel.

A bearded old man in an ornate cloak sitting in a deer-pulled fancy carriage and holding a clock with sun, moon, and the zodiac behind him is the allegorical figure Time in the tapestry *"Time's Triumph over Fame".* This tapestry is one in a series of six tapestries based on Petrarch's allegorical "I Triomphi," a poem about the triumph of love over mankind, chastity over love, death over chastity, fame over death, time over fame, eternity over all. In another tapestry in this set, Aristotle, Alexander, Plato — join in a carriage ride pulled by white elephants. The tapestry series, North French or Flemish, early 16th century is in the Metropolitan Museum of Art.

Bordure (border) Is a bordure needed or not? Gothic tapestries have no bordure. Occasionally a band of text finishes the tapestry. Most modern tapestries have no bordure. Tabard likes to weave a space of empty weave around the subject. Lurcat sometimes puts his composition inside an irregular square, oval or other form on a plain black background which frames it. The taste of the 18th and 19th century was, however, different; medieval tapestries without bordure seemed naked and incomplete. A 19th century art expert's opinion was that a tapestry without bordure is like a wood cut without margin, like a beauty without a head. Domjan tapestries have no bordure and the composition goes out to the very edge. Those woven in Madrid have a one-inch black ribbon — selvage — around them.

"The History of David" a tapestry woven in Brusselles during the 16th century has a bordure, a few inches wide, of rose garlands with dasies and grapes, tied with a ribbon. Italian tapestries prefered rich bordures of Renaissance garlands which had a wide influence on French tapestry, as we can see from the tapestry collection of Francis I, King of France, woven at Fontainebleu.

79

Bordures of Raphael's *"Acts of the Apostles"* would require a separate study for their richness of imagination, beauty, and wealth of subjects, surpassing by far all bordures made earlier. In style they fit the Loggie with elements of the grotesque and combination of putti, classical architecture, figures. There are whole scenes, in monochrome, like relief sculpture with scores of figures, mostly horsemen. The most dramatic among them is a whole battle scene: *Cardinal Giovanni de'Medici at the Battle of Ravenna.* Scholars in the Vatican must have prepared the synopsis — iconography even for the bordures, like Faith, Hope, and Charity, or The Hours. Raphael's pupils, Francesco Panni, and Giovanni da Udine made the designs for some of the bordures, but the bordure we are most interested in, *The Fates,* was made by Raphael himself. It is the vertical part of the border; on top of it is the symbol of Papacy, the tiara and the keys of St. Peter. The Medici coat of arms, six balls of wool, are surrounded by laurel branches and putti holding snakes. The lovely young woman figure of the first Fate sits on top of a vase, or fountain-like device in contra posto, spinning. The second Fate underneath is twining thread on a bobbin; putti, grapes, garlands of flowers and a lyre enrich the design. The third Fate in the rich draperies of classical costume sits under a classical arch, she has scissors and cuts the thread. Two horned old fauns on the bottom are holding all of this on a ball. This ornamentation is unquestionably the triumph of decorative arts. It must have exasperated the weavers even more than the psychological expressions on the faces of the main characters; nothing like this had ever been asked from them before. Raphael did not take in consideration the limits of weaving.

Le Brun was inspired by Raphael. In his series of *The Royal Palaces* he uses rich and complicated bordures; a bordure of leaf scrolls with flowers, the coats of arms of the King in the upper middle, then an architectural frame with marble pillars and columns as if we would look upon the scene across loggias or a stage decor, with ballustrades, festooned with garlands, of fruits. There is the subject *"Le Chateau de Chambord"* at Les Gobelins (1668-81). The popular Chateaux series has been repeated several times.

Brusselles continued to produce tapestries in Raphael's style, large figures with wide and elaborate bordures of scrolls, masks, urns, vases, trellises; the lower middle has often a collection of arms and armour, complete with flags, coat of arms, generous style lifes of fruits, dogs, pheasants and peacocks.

A factory-like mass production of stereotype bordures from old cartoons continued during the eighteenth and nineteeth century.

Woven frames created the fake illusion of heavy carved gold or stucco building elements, with shadows, tromp d'oeil. Once the idea started there was no limit to its use. It was carried to absurdity. A whole Gobelin tapestry handled as a wall, stucco frames, patterned wall paper and in the middle a painting is hanging from a string, framed in a gold frame — all woven in one tapestry. At that point tapestry as such disappeared; its beauty, force, candour were corrupted; the weaver was reduced to a mechanical imitator of a cartoon that doesn't merit realization.

Tapestry by the year 1736 was a lost art.

The Grotesque. Vasari has this to say about the grotesque or capricious: "A kind of painting full of license and absurdity admired by the ancients as room ornaments. Not subject to ordinary rules. For example a weight is attached to a very slender thread which cannot possibly support it, or leaves sprout from a horse's leg — and the more strange the imaginative license, the better it is to be."

The grotesque is an agglomeration of the extravagant, the bizarre and the absurd; it has variety and invention. It is used to amuse like the comedy and the farce in early theater.

Frescoes of Pompeii at first followed architecture as a decoration of the wall. Later refined tastes required more than the enclosed wall. Then frescoes opened up views of marvelous landscaped gardens and fanciful fountains, views of the sea with villas on the shore in perspective, framed with tromp-d'oeil curtains. At a point of over-ripe civilization, even this did not satisfy the restless taste; the painting had to produce what architecture could not do — the impossible; rows after rows of columns and the fantastic Rococo decorations of Pompeii, columns resting on a flower garland, a heavy sculptured bust hanging on a ribbon held by a little bird in flight.

Vasari writes in the life of Giovanni da Udine: "Digging in the ruins and remains of the Palace of Titus, in the hope of finding figures, certain rooms were discovered completely buried under the ground, which were full of little grotesques, small figures, scenes, with other ornaments of stucco in low-relief. When Giovanni accompanied Raffael, who was taken to see them, they were struck with amazement, both the one and the other, at the freshness, beauty and excellence of those works. These grotesques with their delicate ornaments of stucco divided by various fields of color and with their little scenes so pleasing and beautiful, entered deeply into the heart and mind of Giovanni . . ."

Revival of classical ornaments brought a new feature to Renaissance palaces. Loggias of the Vatican and the Uffizi corridors are painted with grotesques; one marvels at their richness and seldom stops to analyze the many fine medallions and details. There are a few neglected tapestries in the corridor of the Uffizi with grotesques designed by Ubertini and woven in Florence in the workshop of Karcher about 1550. These tapestries are wool and silk, with gold, and silver thread. Columns, maidens, scrolls, drapery, busts, putti, garlands of fruit and flowers are arranged with playful fantasy. While most tapestries of the Renaissance were great compositions, full of figures, these were even less appreciated during the following centuries.

The grotesque has no story to tell, no moral to teach—it is non-figurative, even though it has elements of figures, animals, shells, existing things, put together in a way not found in nature. They are used as ornament. While complicated literary subject is alien to modern art, the grotesque—with its touch of surrealism and its only raison d'être to pleasure the eye—awaits rediscovery.

Medieval manuscript illuminations had their elements of the absurd, and text margins had painted decorations in grotesque style centuries before the Renaissance of classical influence. Jean Pucelle's *Belleville Breviary* made in 1323-1326, and now in the National Library in Paris, had for example figures with musical instruments on one page with an iris, a dragon fly and a butterfly larger than a monkey who sits on a trellis.

On margins of the Visitation page of the *Tres Riches Heures* of John, Duke of Berry, we can see grotesque designs; in this case the grotesque can also be interpreted as comic or caricature. There is a fortress tower rising above foliage, behind its ramparts, a knight, fully armored, his visor pulled down, is shown waging battle with a lance against a snail. Another snail comes climbing up on the frame of the page.

King Francis I of France founded the tapestry factory (in operation until 1590) at Fontainebleau in 1530. The sixteenth century was a period of change, novelty, innovation in tapestry making. In the new Galerie de Fontainebleau, the walls were painted with frescoes in the Italian style and framed in elaborate stucco work. Pagan goddesses wandered among trophies, masks, scrolls, medallions in grotesque ornaments. The King ordered tapestry replicas of his favorite walls woven to carry with him on his voyages. The compositions were intended for wall decor, not woven pictures. Maidens—fleshy nudes—emerging from masonry volumes, are holding garlands; plain background modeled after antiquity and heavy frames ushered in a new style in tapestries.

Several large decorative interiors were preserved in tapestry copies, although the originals were destroyed. Louvois ordered a reproduction of Miquard castle at St. Cloud from Les Gobelins.

Grotesque — Classic tastes of the high Rennaissance in Italy. Raphael's followers, Bronzino, Giovanni da Udine, and Bechiecca further developed Grotesque ornament and had a great influence on Brusselles tapestry bordures. Brusselles workshops had orders from Italian princes; the style was so popular all over Europe that cartoon artists had no other choice than to identify with Italian high Renaissance. The bordures became wider and wider and at one point the picture disappeared in the bordure—the whole tapestry became grotesque ornament. Pieter Coecke van Aeslt, the first to design such a tapestry, was a woodcut artist; in his magnificent bookcovers, and title pages, he used every weapon of the arsenal of classical ornament. His clear linear grotesque ornaments are of architectural elements in strongest symmetry and flowers, plants are almost non-existent. His cartoons were used in Brusselles; he was followed by others. Grotesque tapestries, like the *Monkey tapestries* in Madrid, woven initials S.B. Cornelis Floris, a designer from Antwerp, was one of the followers. An ornamentalist in Grotesque, he had strangely eclectic designs; a wide border of animals in forest and swamp background, water birds on the lower part, singing birds and predators on the top of the bordure, cats, quadrupeds on both sides. The animals are naturalistic and woven with superior craftsmanship—each feather as on the verdures, hunting tapestries. This is just one border; then comes a grotesque border of scrolls, tassels, drapery in abstract trellis and another frame with a small picture in the middle. The animal borders—(did, by any chance, some Indian miniatures reach Bruxcelles?)—were repeated and imitated in several tapestries of the Wilhelm Pannemakers workshop in Germany. In 1562 Philip of Spain bought the eight piece series of the *Apocalypse*. Grotesque tapestries continued in Bruxcelles and when we compare one of the early 17th century tapestries with Ubertini's, we can see how the elegant airy composition became heavy and crowded so that no background space is left open and when we think that Ubertini's ideal was a Roman frescoed villa we have a feeling that the harder it was to follow, the further it got from the ideal.

Great Art Patrons

Guilds of Florence

Lady of the Castle

Four Great Art Patrons. Great are the times in history when art is appreciated and understood; when a need for creativity is felt and satisfied by talented men. Among the great art patrons of history were four brothers who lived in the 14th century: Charles V, King of France, a great collector of tapestries and illuminated manuscripts; Jean, Duc de Berry who commissioned the *Très Riches Heures;* Louis, Duc d'Anjou, who commissioned the *Angers Apocalypse;* and Philip (the Bold) from whom the *Battle of Roosebeke* was woven. The brothers competed with each other to possess the richest collection of tapestries; and we can say that all four men were winners.

Charles who reigned from 1364 to 80 was called "wise artist knowing architect" by his biographer. He had a great taste for precious objects and surrounded himself with magnificent works of art. The luxury of his court greatly impressed foreign ambassadors and dignitaries on state occasions. His collections were housed in the Louvre and in the Dungeon of Vicennes. His great collection of books and manuscripts became the foundation of the Bibliotheque National in Paris.

The King built the nine towers of the fortress castle of Vicennes and completed the monumental dungeon in the center of the large building complex. The dungeon still stands today, but the towers no longer exist. Lower floors had apartments for the Royal family and for servants. Later, political prisoners occupied the rooms; Marie Antoinette, among others, was a prisoner here. Climbing up the hundreds of steep steps of the narrow, winding staircase to the tower, one has a wide view of the outskirts of Paris at a place that used to be the forest of oaks and hunting grounds of many a French King.

The December page of *Les Très Riches Heures* du Duc de Berry, painted by the Limbourg brothers, shows a wild boar hunt in the forest of Vicennes. A pack of dogs throw themselves on the fallen boar to tear it up. Their body movements are beautifully pictured as a result of keen observation. More so, they are not drawn separately but in a confused and lively group; yet each dog can be identified according to race and characteristics. The Duc loved dogs and we have a good reason to believe that the dogs are a group portrait of his favorite ones.

On the left is a tired hunter with a spear and on the other side another hunter blows his horn; the hunt is over. Around are the oaks, the foliage in shades of ocher, and fallen leaves on the ground show late autumn. Again the trees are not standing separate but the depth of the forest is painted, the first time in history. Above the forest far behind we see the square towers and in the middle rises the later dungeon. When the miniature was painted the buildings were not finished yet. At the time of their completion Charles V deposited here his collections of illuminated manuscripts, tapestries that were safeguarded in this magnificent stone vault.

Jean Fouquet (born 1420) also painted a picture of the dungeon in a miniature. Fouquet, native of Tours, was a panel painter as well as miniator, the first great French Renaissance artist, he was trained in Paris. He painted portraits of the Kings and Pope, he designed the King's tomb and organized festivities, "Peintre du Roi" (Louis XI) from 1474. He painted views of Paris with great accuracy in his *Book of Hours of Etienne Chevalier* during 1450-55.

Philip the Bold, Duc of Burgundy, was the founder of the Charterhouse of Champmol 1383; his palace in Dijon is the Musée des Beau Arts today with many art objects and fragments from his collection. Master-weaver Jacques Dourdin, citizen of Arras (active about 1380-1407) wove *Le Roman de la Rose* tapestries for the Duc of Burgundy, also pastorale and hunting scenes. Master-weaver Jehan Cosset made the *History of Alexander* for the Duc, who, like his brothers had one of the most important collections of tapestries; in 1382 the Duc ordered from Michel Bernard, master-weaver of Arras, the largest size tapestry to be made, the *Battle of Rosebeke*. It was 220 feet long — even with a retinue of servants it was impossible to lift or hang this size of tapestry and so the Duc ordered it to be cut into three pieces in 1402; even then it was much too large and was later cut again.

Louis, Duc d'Anjou, had a great passion for tapestries and was one of the great patrons of history; he ordered the *Apocalypse* for the fortress of Angers. He was a personal friend of the artists; he followed the development of the work; he was generous. These were the times when artists were paid a bag of gold.

The large room of the hospital had rows of beds for the common people and the beds were divided by tapestry curtains with the Duc's and his wife's coat of arms. This was not the only occasion that tapestries were used in hospitals: invalids' beds in the great ward for the poor at Beaune had tapestries made for Chancellor Rolin with initials, coat of arms, branches and facing birds. Imagine the silent nuns and the marvelous colorful tapestries. After a hundred years of sterile white and drab surroundings there are new experiments to paint hospital rooms, corridors in brightest colors and put back the art work where it can do so much good.

The four brothers are forever immortal for the art work they commissioned. Their wars, their political activities are recorded, but it is through their collections of art that their names are remembered.

The miniature was the source that swelled into a monumental flood, mural painting, sculpture, stained glass and tapestry. Clunaic monks were to spend their lives in noble occupations: praying, meditating, singing, and the cultivation of certain selected arts and crafts such as painting, casting bells, candle making, woodcarving. There were large rooms, scriptoriums, where monks worked on illuminated manuscripts.

Book of Hours were personal religious manuscripts, private prayer books beautifully illuminated page after page with fine, detailed paintings. They served for prayer and meditation as art enjoyment and were popular in France during the 14th century. Like the small folding altars of Italy, the Book of Hours traveled with the owner as his most precious object. While the ownership of earlier Book of Hours was usually indicated by initials, inclusion of coat of arms, in *Les Très Riches Heures* the patron — owner, Jean, Duc de Berry becomes part of the book, serves as model in some of the compositions, has his portraiture painted, members of his household are depicted; his dogs, his chateaux and precious objects, like his favorite tapestry figure in the book.

Les Très Riches Heures contains the famous calendar pictures of the Limbourg brothers; texts for each liturgical hour of the day; Hours of the Cross; prayers, masses. There are large, full page pictures in the book not medallions in the text. Several large painting pages were added like the Anatomical Man; Plan of Rome; showing the Duke's many sided scientific interest.

The Limbourg brothers, Paul, Herman and Jannequin were natives of Nimwegen, Germany. They were apprentices at a goldsmith in Paris. Imprisoned in Bruxcelles in the political turmoil, they found employment at Jean, Duc de Berry. It is certain that at least Paul worked in Italy, Milan and Siena. The brothers worked for years for the Duc, and attained higher and higher levels of perfection. The artists and the patron exchanged gifts each year in January. Their intimate relationship permitted the artists to play a joke. In 1411 they presented the Duc with a fake book. A piece of wood was beautifully bound but had no pages nor text. This book is in the Duc's inventory in all seriousness. In January, 1412, the Duc presented to Paul Limbourg a diamond-and-gold ring. He also got an emerald ring from the Duc. In August, 1415, 1000 gold ecus were given to Paul as payment. Following January Paul gave the Duc an agathe salt cellar with gold, sapphire and pearl decoration. He received a ruby ring. Paul was made "valet de chambre" to the Duc and head of the workshop.

Jean, Duc de Berry was born in the Chateau de Vicennes. As the son, (Jean II the good, King of France reigned 1350-64) brother (Charles V, reigned 1364-80) and uncle (Charles VI reigned 1380-1422) of consequent kings of France he was forced into politics.

The Duc owned large estates, 115 castles, some of which are pictured in *Les Très Riches Heures.* He was an extravagant, passionate collector of art, he was a generous patron of the artists, running into debts to pay for the luxurious treasures. The castles were filled with precious objects, books, gems; his tapestries traveled with him and adorned the walls of banquet halls where he organized festivities for his friends.

The Duke owned at least twenty exceptional rubies, jewels from Italy as well as Luccan gold brocades, silk hangings, gold plates, forks (an innovation!) He also loved exotic animals and had ostriches, camels, a bear, 500 dogs. His collection of illuminated manuscripts was not as large as that of the King, Charles V, his brother, but he owned books of exceptional quality. He took care that his books were bound in the most luxurious bindings; silk, velvet, leather, embossed, gold and silver fastenings, clasps decorated with enameling and carbouchons.

During the dramatic events of the times his Paris residence Hotel de Nesle was pilfered and the Chateau de Bicetre, outside of Paris was burned. Paul Limbourg's paintings were lost in the fire. After the loss at Agin in the war against the English those close to his heart perished and he died in distress.

The month of January is the only interior scene in the *Très Riches Heures,* a group portrait at a banquet table. The Duc is there in a fur hat and brocaded gown; this is the occasion of the exchange of gifts. He is talking to the bishop of Chartres, his friend (also a great collector.) The master of ceremonies invites the guests to come closer; Paul Limbourg painted here his self-portrait in a grey cap pulled over his ear, here is his wife and his brother drinking. Members of the court in elegant fashionable gowns warm their hands at the monumental fireplace behind the Duc. The table is set with golden plates. The Duc's favorite pommerianian dogs take a bite from this plate or another. A huge golden boat-shaped salt cellar was evidently made on the Duc's order as it has the bear and wounded swan, heraldic motifs of the Duc's which are also on the red silk canopy behind him, with the gold fleurs de lis on blue background.

With princely carelessness a large and magnificent tapestry is hung on the wall, covering the whole wall behind the fireplace and continuing on the adjacent wall and folded over the fireplace. The tapestry of historical subject, shows the *War of Beque de Belin,* a battle scene takes place, knights on horseback emerge from a fortress, another castle in the background, several groups are engaged in the battle, lettering above explains the scene. The tapestry, now lost, is from Arras, similar to a fragment of *l'Ofrande du Coeur* in the Cluny Museum.

On the page of the Office of the Dead the illustration is placed in an arched church window. The text is only a few lines, small medallions, each a separate little picture, a composition with many figures, a bas-de-page (end-picture) of the story of the three young men and the three dead, fill up the page. The space between pictures is decorated with acanthus leaves in different color flowers, carnation, iris, columbine, forget-me-nots, and birds; a heron, a crane, a pheasant. The medallion in the upper right corner has Death riding a unicorn. The Limbourgs had left this page unfinished and it was completed by Jean Colombe.

Les Très Riches Heures are painted on vellum—parchment. The colors are mineral and vegetable colors ground to a fine powder with a muler on a marble slab by the artist or apprentices. The water base is thickened with arabic or tragacanth gum. About a dozen colors are used in lighter or darker shades. The most expensive color was Oriental lapis-lazuli; the enchanting angels, blue skies are painted with it; in the January scene, the Duc's brocaded gown is painted with lapis-lazuli. Precious was also the light green color derived from malachite; the green fields, the gowns in the May picture, green for spring festival, are made with this color. Other colors were red ochre, a mineral color, vermillion, the brightest red gotten from mercuric sulfide; another red was made of the red oxide of lead; pink color (rose de Paris) was a decoction of red dyewood; yellows a monoxide of lead; violet from sunflower; white was lead ore; black was made of soot. Gold and silver were generously used. Gold was available in two forms, gold leaf and powder; the gold leaf is partly covered at places and the gold powder was mixed into color pigments.

Famous Cities of Tapestry

Beauvais

LES GOBELINS

Les Gobelins

Angers. High above the river banks of the Maine river stands the forbidding fortress of Angers, one of the oldest chateau of the Loire valley. Its immense round towers, built of dark slate, sandstone, and granite stand unchanged for centuries. The massive castle is encircled by boulevards in pentagon shape and by a deep moat. The moss-covered black stones witnessed seven hundred years of history; first as a fortress, Angers was a great achievement of 12th century military architecture which withstood many wars and sieges; later Angers was the setting of pompous courtly life; then it served successively as military post, armory, and finally has become a relic and a museum.

We cross the drawbridge, pass through the vaulted entrance and inner courtyard and arrive at the museum building. Here we find the famous series of tapestries illustrating the Apolcalypse, made for Louis I of Anjou by Nicolas Bataille, master weaver, during the late 14th century.

In ancient times the site was inhabited by the Gallic tribe of the Andegavi. The Romans had a thriving city here with amphitheater, circus, and public baths. Many of the churches of Angers have partly Roman foundations, as well as Carolingian, and Merovingian (pre-Romanesque) edifices, and there are Gothic buildings with the characteristic "Angevin" vault. During the 9th century Angers became the seat of the counts of Anjou. The cathedral of St. Maurice was built during the 12th and 13th centuries with three towers, an unusually High Gothic west facade, stained glass windows dating from the 12th to the 16th century, and a fine collection of tapestries.

Lurcat discovered Angers in 1937. He was looking at the Apocalypse tapestries in the then neglected halls covered with spider webs and dust and read the prophetic message . . . the great drama of the immediate future. Soon the second World War brought forth the scenes of the Apocalypse; fire came from the sky, fire mixed with blood; boats burned in the sea; there was great combat in the sky, fire, blood, and pain on earth, on sea, and in the skies. Men spent their nights in a hell of fear. Great cities were burning: Varso, Rotterdam, Belgrade, Budapest . . . Slavery, famine, and much suffering followed; and then came Hiroshima . . . the seven sufferings, the four routes of terror; the Apocalypse of modern men.

The tapestry series, this most important treasure of France, had lain buried, forgotten, and mutilated for two centuries. "Que diable" exclaimed Lurcat, "Why, the twentieth century can't offer a work of such capacity." He was fascinated with its monumentality — first with its size then with its simplicity. Lurcat became obsessed with the idea of creating a work of similar scope. He studied, measured; calculated the thickness of wool, weaving methods, costs, size, time. Thus started the idea of modern tapestry.

When Domjan was drawn to visit Angers, he already had done a small number of tapestries. Seeing the Apocalypse tapestries convinced him that the route he had taken in tapestries was the right one: the dominant pure colors in a limited palette, the large forms and fine details, the role of ornament.

Face to face with this great tapestry series, we were lost in them. We spent days discovering new details. The fire; the Angel telling about the fall of Babylon; houses collapsing, destruction, burning buildings. Only a few days later in the flaming ruins of our own home and studio, with its lifetime of work lost, the images of the Apocalypse returned to mind and we asked ourselves, did we bring the seed of disaster with us . . . ?

Nicolas Bataille and Hannequin de Bruges. The Angers Apocalypse tapestry is an immense work 5½ meters high by 144 meters long; 800 square meters were woven in a few years. The tapestries are of exceptional quality achieved by great simplicity, "a glory to the artists who created it, to the artisans who executed it," and, we can add, to the patron who ordered it.

The painter was Hannequin de Bruges, also called Jean de Bandol (Hannequin is a nickname for Jean), a well known artist and court painter to Charles V, King of France. On January 31, 1377, 50 francs were paid to Hannequin de Bruges for "portrait images" of the Apocalypse, according to the bookkeeping records of the Anjou. In April of the same year, 1000 francs were paid to Nicolas Bataille for the weaving of two tapestries of the history of the Apocalypse.

Hannequin de Bruges has been compared to Fouquet; the "portrait images" or cartoons for the Apocalypse series were painted with great delicacy, fine modeling, and a unity of composition. The artist had a great idea, the talent to realize it, and the perseverance to go through with it during the years of weaving work. There must have been a perfect co-operation between the artist and the master weaver. They both realized that this large size could not be woven on one loom. Instead, each large panel was composed of separate units, with alternating backgrounds of red and blue, and sewn together.

The Apocalypse was a favorite theme of the Middle Ages. We know of at least sixty illuminated manuscripts of the Apocalypse, the earliest ones come from Spain. In 1373 Louis I of Anjou, borrowed a manuscript of the Apocalypse from his brother, Charles V, King of France. In 1380 the King died; in an inventory taken at this time there is a note stating that the King lent the Apocalypse manuscript to Louis of Anjou. This manuscript served as a model for Hannequin de Bruges. (The Saint Victor manuscript of the Apocalypse, now in the Bibliotheque Nationale, was another source). De Bruges may have seen other manuscripts of the Apocalypse now in the libraries of Namour and Metz, but did not copy any of them. His monumental composition is original.

It was customary for small designs to be enlarged by the craftsmen specialists who made the cartoons. The cartoons had the outlines traced and the color was left to the weavers to fill in. In the making of the Apocalypse, however, the artist probably made the cartoons and planned the colors. He stayed on and followed the work and when he saw that the plain unbroken backgrounds could be enriched, he put ornaments and more motifs in the later panels. He became more daring with his figure compositions from purely decorative Byzantine mandorla to complex scenes where he now painted the multitude.

Nicolas Bataille, merchant-tapissier, head of the workshop, master weaver, citizen of Paris, with the title "valet de chambre" delivered such a number of tapestries to different patrons through his years of activity that it seems he must have supervised several workshops. We have plenty of documents on his work and payments made to him. Nicolas Bataille was a powerful member of the weavers' guild. He used his own seal on parchment documents like noblemen; his seal, round in shape, shows a woman holding a shield with three balls of wool. He married twice and when he died at the age of 60 he left his widow and small children well provided for.

The year, 1373, is the first date when Nicolas Bataille is listed on the records of the Anjou treasury; he received 20 francs for 6 tapestries with coats of arms. In 1375 he received a 20-francs tip for vine; records of the same year show high sums paid to him for tapestries of personages. The description is hazy, so we are not sure if these were in payment for the Apocalypse series. In June, 1379 he became "valet de chambre," a confidant servant to the Duke of Anjou. One of his many functions was to pay other servants. Bataille received 35 francs, of which he was to keep 20 and give the rest to the other valets as tips.

In 1376 Nicolas Bataille delivered 78 tapestries in color for Amadee VI, Count of Savoy. There were heraldic symbols with eagles and knots. In the same year he was paid 1,000 francs for a very large high warp tapestry, *Hector*, made for Louis d'Anjou. He also made for Louis I d'Anjou the *Life of the Virgin*, a fragment of which is now in the Musee Royaux d'Art et d'Histoire in Brussels, and received 300 francs for it. In the meantime the immense work of the Apocalypse series was in progress on the looms in Angers.

April 7, 1377, Nicolas Bataille received 1,000 francs for the *Angers Apocalypse* tapestries. Christmas, 1379, Nicolas Bataille received the enormous price of 3,000 francs for three finished tapestries of the *Apocalypse*. The series was woven probably from 1373-1380, or longer. Each piece must have been of the same size; 7 identical large compositions each with a large personage on the left and 14 smaller compositions in two rows. The large personage is a vertical format while the others are horizontal compositions. A narrow band frames the pieces.

The first *Apocalypse* tapestries have the initials LM, Louis I d'Anjou and Marie de Bretagne, with the coat of arms of Anjou and Bretagne. The next pieces have the initial Y — is it for Yolande d'Aragon, who married Louis II in 1400? We don't know.

Louis I d'Anjou, the great patron, died in 1387. Nicolas Bataille's reputation was well established by this time and he became weaver for the King. During the next thirteen years, until he died, he delivered some 250 tapestries to the royal household, to the Queen, and to the princesses. Charles VI reveled in lavish decorations and his emphasis was merely on quantities. Each door, window, bed, and bench in the royal bedrooms were decorated with tapestries in his medieval Versailles. These had coat of arms, the fleur de lis, verdure and the word "jamais" woven. The King owned sumptuous furniture; gold vessels and jewelry were made for him by his goldsmiths; vases, coins, gowns were made of the richest fabrics; hats, gloves that were gold embroidered and studded with precious stones. His collections testify to his intense passion for luxury and extravagance. Some of these treasures are in the Louvre. We know about the rest through the inventories. Nicolas Bataille received 100 francs monthly in the service of the King.

May 1, 1389, the Abbey of St. Denis was the scene of a fabulous celebration given by the King on the occasion of the admittance of his young cousin, Louis II d'Anjou to the order of knights. Twenty-two knights and twenty-two beautiful virgins selected from the high aristocracy came in procession, and were followed by three days of tournaments. Although scandalous orgies followed the religious ceremonies, the King wished them to be long remembered and ordered Nicolas Bataille to weave a ten-piece tapestry series as a souvenir of the festivities. Nicolas Bataille started it and Dourdin finished it. In April, 1400, Dourdin and Bataille's widow went to the King to receive the money due to her deceased husband.

Paris tapestry workshops were closed in 1418 due to the loss of capital to the English during the Hundred Years War.

The Angers Apocalypse tapestries, Nicolas Bataille's main work, was woven of the best wool of Arras in a coarse stitch with 12 warp threads per inch. Less than 20 colors were used. The original dark blue of the backgrounds has faded to a lighter shade, the red of the backgrounds has faded to a fine rose, and yellow and green faded to many shades of pale grey.

The tapestries were made for the castle of Angers, but probably had no permanent place as there was no one single room of such size that would have accommodated the whole series at one time. Instead, some of the pieces were put into the great hall. Others adorned the castle chapel. They may have been hung outdoors during festivities. In 1400, when Louis II was married to Yolanda d'Aragon, the tapestries decorated the wedding in Arles. King René inherited the tapestries and greatly appreciated his grandfather's treasures; when, in 1458, the castle went through extensive remodeling, the tapestries were put away and not brought out again until 1470. In 1474 King René, in his testament, donated the tapestries to Angers Cathedral; he did not want them to go to Paris. King René died in July, 1480 and in September of the same year, King Louis XI entered Angers. For his reception the Apocalypse series were used to adorn the Cathedral, and were left there until the Revolution.

During the 18th century the Cathedral was modernized. The taste had changed. Charming sweet Pastoral scenes were then the mode. The sincere simplicity of the Apocalypse was no longer understood. The tapestry series was discarded as out of fashion from this time until their rediscovery. The series underwent such adversities that it is astonishing that there is anything left. In the stables of the bishop's residence the partitions were covered by pieces of tapestry to protect the horses from getting bruised. Other pieces served as rugs in bedrooms of servants, melons were covered by pieces of tapestry to ripen faster; when the ceiling was painted the floors were covered by the tapestries to keep them free of paint. No one protested when, in 1843, the one-time pride of kings, the tapestries, were offered for sale. There were no customers, no one cared until the Bishop, Angebault, bought them for 300 francs; he had a sense of their value even in their dilapidated condition. The custodian of the Cathedral, Joubert, was commissioned to direct the restorations. Joubert started his work with great enthusiasm. There were only 48 pieces remaining out of 105, and these ruined fragments were in deplorable condition. Ten more subjects were soon discovered and added to the collection; today there are 67 and 3 fragments. A tapestry workshop was founded to undertake the restoration. Although this workshop was often criticized, without their work the tapestries would have perished. In 1858 Bishop Angebault wrote: "the poor Moses, Joubert took care of him, and like the daughter of the pharaoh, saved him" . . .

The *Angers Apocalypse* series regained their place as highly valued art treasures. They occupied the place of honor in the Cathedral; they were used as superb decor during processions; they were put into the large hall of the Bishopry that became the diocese museum, and after further restorations, returned to the Castle of Angers, where they are now being exhibited.

At a time when the inner fire of faith and a vivid imagination made angels and devils that populated heaven and hell appear more real than any being of flesh, the Apocalypse was a favorite topic.

Probably the *Angers Apocalypse* was not the first time this subject was used in tapestry. The Abbot of St. Florent in Samour had hangings made of the 24 elders of the Apocalypse with musical instruments, after the revelations of St. John. The protective father image, the bearded patriarch, was as much the image of the times as the teenager is the focus of today. Medieval pilgrims, who made the long journey from the Tour St. Jacques in Paris through Chartres to Santiago de Compostela in Spain to visit the miraculous tomb of the Apostle James, were led to the Portico de la Gloria, entrance to the Cathedral. Christ in Glory, richly carved in stone, is in the center, the twelve apostles around him and in a half circle the 24 elders of the Apocalypse, with musical instruments, in praise of God.

Each of the seven large tapestries of the *Angers Apocalypse* series (we have four of them, three are lost) has a large figure of an old man. The old men are probably symbols of the seven churches of Asia, seven bishops of great importance, shown in their churches. We have seen before how literature had an influence on subject matter. There was a novel of seven wise old men, counselors of the Emperor Vespasian (another tale of antiquity like Hector and Alexander) and it may very well be that the old men of the tapestries are taken from the novel of Vespasian. The old men, with long white beards and hair and long tunics, are shown under a canopy of imaginary Gothic architecture holding a book. On one of the tapestries, two red-winged angels on top are holding flags with fleur de lis. Repeated butterflies populate the background. It was also suggested that the large personages are each a portrait of the donor, Louis I. Numbering of the small compositions was changed in our day; there is still a debate on the sequence of scenes, after several changes there may yet be errors, since not one of the large pieces came to us untouched. Each was taken apart and Joubert was looking for matching pairs rather than following the sequence of the subject matter.

On one of the small scenes 24 old men are seen in adoration. Christ in a mandorla is in the center. The old men put their crowns at his feet. Individual faces, different hair-dos, beards — they all are delicately made. St. John is present in each scene; and in each he has a different attitude. He is the composed spectator standing in a doorway, he is part of a crowd, he is looking out of a tower window, bewildered, his finger in his mouth. He is distressed at the sad happenings, dries a tear discreetly on his sleeve, he is involved in the scene and in another he is the main character. He sits on a rock, writing his book and is surprised by an angel. St. John steps out of his house and the angel takes him by hand, to show him the great prostitute. The great prostitute is a meager maiden sitting on a rock, combing her hair, a rare and daring presentation of a half nude.

95

Luis XIV   Donjau

In the tapestries, nature, clouds, trees, mountains are not observed but stylized. Often the influence of Byzantine miniature painting can be seen. Clouds are ruffles, repeated. Trees are smaller than men, there is no perspective; this two dimensionality fits to the tapestry beautifully. Small flowers and animals on the edges are more naturalistic.

The Apocalypse abounds in elements of the fantastic; and the artist followed the text word by word; man-faced war-horses have a dragon's tail. Enormous insects attack frightened people; a king has wings of a bat; there are horses with lions' head and snail-tail; monsters-heads look out of the clouds; Satan, dragons, beasts — all have 7 heads. The fatal number seven is repeated through the composition; 7 churches of Asia or 7 bishops; they have 14 angels; seven gold chandeliers; seven seals; seven trumpets; seven gold cups; seven kings; and seven mountains.

The backgrounds are plain at first; later ones are patterned; evenly placed repeated identical flower motifs in white on blue background; irregular vines; green leaf scroll pattern connects the figures; checkered pattern gives a peaceful background to the dramatic scenes of St. Michael's fight. Repeated initials in blue and white on red background are modern, daring. The basic red and blue background colors give unity while the variety of ornamentation in the background gives change and interest. The ornament is eternal — if thrown out of the door, it comes back through the window and pops up unexpectedly in cubism, Matisse, Vasarelly, the triumphant ornament — art cannot be without it.

Versailles: King Louis XIV was a youth of twenty-three when, in 1661 he decided to embellish his simple little hunting rendezvous place, the Chateau de Versailles; with the gardener, Le Notre, the architect Le Vau, and the painter Le Brun.

In February 1653, at a ballet at the Palais Petit-Bourbon, Louis XIV appeared in his Sun costume. In its final form the Chateau de Versailles became a hymn to the sun as symbol of the King; "Le chateau est un hymne au soleil royal." The legends of the sun in mythology; Marsyas, Daphne, the Cyclops are present. Each royal room has for its subject a planet that circles the sun allegorically represented which refers to the King. Zodiac signes are used frequently. In the room of the Queen the sun, a radiant head of Apollo, symbol of the Sun-King, lights up the four corners of the earth. The sun rises over the basin of Apollo, the King wakes up and the morning ceremony starts; the dressing of the King in the presence of privileged aristocrats and the greeting of the King on his way to the palace chapel. Versailles reflects the King's love of space and light. In a country of absolute monarchy, there were no more enemies. Before windows were small to insure protection — openings were for defence. Now the big windows were invented. An endless row of rooms, magnificent guilded door frames, one after the other in perspective conveys the feeling of grandeur. The mirrors reflect and multiply the apartments on one side and the gardens on the other — the gardens prolong the system of symmetry. Frescoed ceilings open up into a blue sky. Gold dominates; guilded frames and furniture, golden brocades, gold thread in the tapestries. Many were destroyed during the Revolution for their gold contents alone.

The building of Versailles, the decoration of Les Grands Apartments, the planning of the Parks with water-works, monumental fountains, vases and statues, the pomp and splendor that surrounded the Sun King, and which, with other excesses, brought about the bankrupcy of a nation, employed hundreds of artists and designers who turned out technically perfect works in the pompous academic style. Le Brun alone reigned over the best artists of the country and had 800 tapestry weavers at the Royal Workshops of Les Gobelins.

Les Gobelins. Gobelin means tapestry and the word is also used, mistakenly, for needle-work. Tapestry is woven on a loom; needle-work is made with a needle and yarn on prepared textile.

The Gobelins were a family of dyers who probably came from Rheims during the fifteenth century and settled on what at that time was the rural countryside, outside of Paris, left bank, at the Bièvre river. Jehan Gobelin, discoverer of a scarlet dyestuff, died in 1476.

King Francis I started tapestry workshops at Les Gobelins. King Henry IV was very much interested in tapestries. He brought two Flemish weavers (Marc de Comans and Francis de la Planche) and put them to work in the Gobelin family house. With their team of low-warp weavers they were producing tapestries in the Flemish style. In 1608 he established a group of weavers in the Large Gallery of the Louvre who worked on upright looms. The Gobelin family grew in wealth and esteem. Balthasar Gobelin had several influential offices at court; among others; councillor secretary of the King, treasurer extraordinary of war, chancellor of the exchequer, etc. He received land from the King, Henry IV, and became Lord of Briecompte. He died in 1603.

The Gobelin workshops returned to upright looms, and were expanding. Colbert, the great statesman, financial genious, and adviser to the Sun King, Louis XIV, was well aware of the great asset the Gobelin workshops meant to French economy. (Ever since tapestry production was and is an important factor in French economy). Colbert purchased the Gobelin workshops in 1662 and in 1667 it became Royal Workshops, one of Colbert's model factories financed by state funds. Colbert placed Charles LeBrun (1619-1690) at the head of the workshops as director. Colbert, himself a collector, was devoted to the promotion of the arts and could find time for it in spite of his many demanding offices. In 1673 he presided over the first publicly held exhibition of paintings by living artists. He started the building of the Louvre, as we know it, and brought Bernini from Rome to be its architect.

LeBrun was born in Paris; at the age of fifteen he already had a painting commissioned by Richelieu; at twenty-three he was in Rome painting with Poussin. With Colbert he founded the Academy of Painting and Sculpture and the French Academy in Rome. LeBrun studied ancient tapestries; he was most interested in the technical part of production, the weaving, the dyeing of wool. He made oil sketches and copied his own painting for the cartoons in oil, on canvas, of his tapestry series.

Tapestry at this time was mostly commemorative, often taken from contemporary history, triumphs of wars, for official propaganda, and diplomatic gifts. LeBrun designed the *Royal Houses* series, the *History of the King* from the life of Louis XIV. His other series include traditional topics, the *Elements*, the *Seasons,* were an eternal topic, Charles V had a series of the *Twelve*

Months, 300 years earlier. LeBrun's series *Alexander* was also a favorite subject; Philip the Bold had a series of *Alexander.* If subjects were traditional, the interpretation was totally different. These were large size tapestries with life-size figures in the foreground, small figures, maybe whole battle scenes in the background, with rich bordures. Chiaroscuro, air distance, perspective, innovations in table painting now became requirements of tapestries.

There were many changes that took place gradually in the method at Les Gobelins. LeBrun insisted that the weavers work from finished cartoons — oil on canvas at that time — and not from an outline of forms with a vague indication of colors to be used. Colors had to be followed exactly. Working on the cartoons were groups of artists who specialized in drapery, architectural elements, landscape, figure, flowers, and animals. The weavers' training began to include classes in drawing and painting. The weavers also specialized in making figures, bordures, and animals; the heads and hands were always made by weavers who had the greatest skill and longest training.

Many new colors and a finer weave were needed. The series, *Life of Louis XIV: The Audience with the Pope's Legate* had one hundred forty-one different shades of forty colors, each in ten varieties. Earlier a red mantle was woven with four to five shades of a red, but now the shadows were deep purple, the highlights almost white, including many more colors and fine transitions for this one area alone. Still, there was moderation, a systematic repetition of the same hues for various objects through the tapestry that assured a unity, a cohesion, later discarded. LeBrun was a perfectionist; if he was not satisfied with the execution of one of his tapestries, he had the part remade.

On October 15, 1667 the King visited the Les Gobelins workshops. This visit became the subject of a tapestry included in the series, The *Life of Louis XIV.* The large 5.7-meter silk and gold tapestry shows the King, around him are Colbert, other nobles; and more come in through the half open door of the workshop. The King is being presented by the artisans of the workshop; by a rolled-up tapestry, a silk curtain just being taken down by an apprentice; a large silver plate is carried by two craftsmen; inlaid tables, marble vases, a painting, textiles. A large tapestry is hung on the wall in the background. The complex composition has a rich bordure.

LeBrun was responsible for interior decoration of the palaces of the King. He designed the Gallery of Apollo in the Louvre; the Halls of War and Peace and he painted the allegorical frescoes on the vaults of the Great Hall of Mirrors in Versailles. Les Gobelins at that time included upholsterers, cabinet makers and an army of fine craftsmen who worked on walnut furniture, tables, cabinets, inlaid with semi-precious stones, marble, tortose shell, foreign woods, ivory, porcelain plaques, mounted with brass, and heavy gilded bronze; "meubles de luxe" for Versailles. We can say LeBrun was the creator of the Louis XIV style, the pompous, rich decoration that pleased the Sun King and was in turn followed and imitated by every monarch; the obligatory pomp of courts, Madrid, Budapest, Naples and Caserta . . .

The sun still rises over the pool of Apollo and hordes of buses arrive — a new day starts in Versailles. Louis XV was celebrated in the tapestry series of the *Hunt* by Jean Baptiste Oudry, who became official painter and director of the Beauvais Royal Workshops and superintendent of Les Gobelins. Oudry came after decades of technical experiments. A new generation of weavers, docile employees of the state, tried to meet growing demands. Oudry asked for point for point imitation; the tapestry now had to create the perfect illusion of an oil painting. In 1736 Oudry circulated this message — "The weaver now has to give his work the spirit and intelligence of paintings — only in this rests the secret of making Gobelin tapestries of first quality." There was a serious conflict between the weavers and the director and the weavers did not give in easily, as we can learn from a letter by the master weaver of Les Gobelins; "Good painting and well woven tapestries are two absolutely different things. There exists in the storerooms of the King's Furnishings tapestries carried out by weavers only. They are, as regards to color, all that tapestries should be; which must need display more vivid hues than do paintings." The weavers blamed the direction for the discoloration in only one two decades of tapestries made with the new enlarged color scale.

In 1671, Colbert set the rule for the master dyer. The questionable so-called minor tints were prohibited. There were 120 colors. This rule was dropped long ago. The most delicate shades were most perishable to light, air, torch or candle light. In 1748 Oudry writes; "The opposition which our most distinguished painters whose pictures have been executed in tapestry of Les Gobelins have for the last thirty years encountered on the part of the workers shows how far we are yet removed from such cultivated taste found on a knowledge of right principles". The gap only widened. Fifty years later Lacordaire is proud of the progress; instead of three normal flesh colors in twenty shades, five hundred and twenty-eight different flesh shades were produced for one tapestry alone.

Hatching was not used long ago; instead color patches imitate brush strokes. The great virtuosity was portraits. Thinner and thinner wool and silk was used with more and more color shades. Pigments were mixed in dyeing, but this was not enough and the very fine thread was of two or three different colors, inferior and perishable dyes. The finesse of weaving arrived to imitate the varnish and cracks of painting in tapestry.

During the Baroque period the rich interior decoration was independent of the severe classical architecture. The Rococo brought a fusion of architecture, decoration, and a unity of proportions and scale. Curtains, walls, had to be coordinated to the last bonboniere. Tapestry rooms now included upholstered chairs of the same design and color.

Gradually there has been a complete change in the ways of use of tapestries; the mobile decoration of the Middle Ages. At that early time on iron poles between columns, on hooks, iron nails on walls, and beams, tapestries were hung for certain occasions, casually, as we have seen in the *Très Riches Heures* of the Duke de Berry. Free-hanging tapestries created a room. Even if at a permanent location tapestries were always "hung" not fixed on four sides till the Baroque, when they were — part of the decor — fixed to the wall with a frame.

The eighteenth century is a great period in French interior architecture; tapestries are woven on measure for one particular room; and to cover each wall there are "entre fenêtres", "end pieces", borders", "portiers" woven; these are stretched and fixed to the wall. Matching sofas complete the room.

Madame Pompadour had her furniture designed by Boucher.

Francois Boucher (1703-1770) was also born in Paris and first worked as an engraver on book illustrations. After years spent in Rome he returned to Paris. In 1755 he became director of Les Gobelins and court painter. He was a favorite artist of the times that wanted to be entertained. His sweet feminine pastorals bought a lighter tone into the strictly classical style of his predecessors. Boucher also supervised the Beauvais tapestry production. His cartoons created new problems for the weavers from a technical point of view; for the many fine pastel shades that were hard to follow. To reproduce Boucher's paintings in Gobelin tapestries, their charm, elegance, and sophistication, the weavers had to learn how to paint with the bobbin.

We must only think of the Angers Apocalypse series of tapestries to see how far the weavers followed the painter. Hundreds of new dyes had to be made both in wool and silk to reach the 10,000 hues required for the fine picturesque effect of the Boucher cartoons. 20-40 warps to the inch were needed for the necessary finesse. Technical standards were very high but from an artistic point of view there was great deterioration. There was a great waste in dyeing that many shades; because of the finesse of weaving, the work took much longer, production costs soared; during the second half of the 19th century even the state could not affort to order large size tapestries.

The Metropolitan Museum of Art installed the *Rose Tapestry Room,* with the original furnishings of Lord Coventry. These had been ordered for Croome Court, his old Worcestershire house which Robert Adam was remodeling. Boucher paintings, in an oval gold frame on the just invented damask-like background, form the center of each tapestry. The tapestries cover the wall surface from edge to edge and from the ceiling to the woodwork. The tapestries have a gold bordure and garlands of flowers; from these the oval paintings are hung by ribbons. The strawberry rose background continues on the sofas and arm chairs decorated with scattered floral arrangements. The rug on the floor reflects the Adam ceiling. The tapestry room was greatly admired and imitated in England.

Chinoiserie Le Roi de la Chine was the title of a ballet given as "divertissement" on a ball in Marly, organized by Louis XIV in January, 1700. This date marks the high point in the interest in the bizarre, the unreal, the exotic, the far away and unknown lands, never to be seen, a nostalgic longing so beautifully expressed in Watteaux's Embarkation for the island of Cythere.

Much earlier, Chinese goods reached Europe in weaves and moved the fantasy of those who could see them; art lovers admired them; artists were inspired by them. By the end of the thirteenth century Chinese silks were brought to Europe with motifs of the imperial dragon, the Fen-huang, and scrolls of clouds. Marco Polo in the thirteenth century, Mendoza in the fourteenth tell about the distant lands of the Far Orient. During the reign of Louis XIII at a fair at St. German, Portugese merchants were selling Oriental goods from the colonies that delighted the public and started the fashion of "chinoiserie". The name "chinoiserie" was applied to small "objects de lux," imported from China or made in the Chinese manner. These objects were complicated, of intricate craftsmanship, and unusual in their shape and use; porcelain, lacquered boxes and screens, furniture, silks. Fans became so fashionable that in 1678 the fanmakers' guild was founded in Paris. Chinese motifs appeared on French silks, porcelain, furniture. The imitations did not aim to be authentic copies, but exaggerated, fanciful versions, Chineser than the Chinese. Chinese vases were placed on the mantel, fruits, desserts were served in Japanese plates; whole Chinese rooms were built; the decor of the Chateau de la Muette in Rococo style and the Chinese rooms of Versailles and Schonbrun. Madame de Pompadour took a personal interest in Chinese fashion. As chief shareholder in the "Compagnie des Indes", she favored the Chinese decoration also in French adaptation and is said to have designed some silks herself for the Lyon silk industry.

The Jardins Chinois tapestry series by Boucher gives a soft pastel touch to the exoticism a la mode. *The Aviary* tapestries, after Boucher were woven in Aubusson, late eighteenth century; of imaginary birds in fantastic cages under palm trees. In Beauvais a set of so-called *Chinese Tapestries* were woven before Oudry. Boucher designed the new, second series of *Chinese Tapestries* for the Beauvais workshops and these became so popular, they were repeated several times from 1743 to 1775. Aubusson also repeated his series. The tapestries are compositions of figures with slanted eyes, in rich draperies, wide sleeved gowns, imaginary headdresses, turbans, straw-hats, holding fans under curved fantastic baldachins. Some negroid characters are always mixed in the crowd, the background buildings are an uncertain mixture of Roman ruins and Chinese pagodas in a rosy haze, partly covered by exotic trees and the whole picture in soft yellows, pink shades, a light cream colored dream world.

It took two years on the trade routes for a Chinese vase to reach France from Peking and although we can fly around the world and return in a day or two, the fascination of the exotic is still with us, with the young men wandering across India or the fashion trend interior decorators always returning to this inexhaustible source.

One of the *Chinese Tapestries* is the one with the *Great Mogul,* who sits under a canopy of silk drapery and tassels, surrounded with figures in Orientalized costumes; there are flags with Chinese insignia, exotic plants — and the coat of arms of the Kurfurst Klemens August of Koln, for whom the tapestry was woven in Beauvais, 1730.

The so-called *Asia Tapestry* by F van der Borght, Cologne, is one of the exotic tapestries woven in Germany; it has a camel with a baldachin, and turbaned figures.

In 1504 already the archduke Philip le Beau bought a series of tapestries of Portuguese-Indian subject; these were large tapestries with giraffes, leopards, elephants, and Indian, exotic fish swimming in the rivers. Another favorite subject was the *Egyptians* — gypsy scenes, the carefree life of the nomad caravans, their colorful costumes. Chinese, gypsies, or Indian, the costumes, plants, buildings were imaginary, the scenes romantic, theatrical.

Flemish Tapestry. The Arras tapestry workshops were so important that tapestry in general was called Arras work, arrazzo. During the fifteenth century tapestry production shifts from Paris to Arras because of the Hundred Years War. Then Arras is invaded by the soldiers of Louis XI and Bruges and Tournai become the centers of weaving, and later Bruxcelles, although there were workshops active simultaneously. Cartoons came from Antwerp painters, local artists or so far as Italy, Raphael's cartoons for the Apostles.

Flemish weavers produced an immense quantity of tapestries, most of them in largest sizes, in series, rich and complex compositions of figures, crisp, clear design, many fine details that asked for perfect craftsmanship. From earliest times Arras was the center of spinning and weaving carried through on an industrial scale. Here for the first time many looms were put into one large room, this was not a home-industry any more. Without a well-organized industry, available good raw materials, wool, silk, metal thread and well-trained skilled weavers it would have been impossible to fill the great demand made to Flemish weavers by their patrons. During the early sixteenth century protesting weavers were forced to emigrate; they went to Italy, Germany, England, and some to Aubusson — and so the art of tapestries was spreading from one country to the other.

The magnificent tapestries of the hunts of the Duke of Devonshire; *Bear and Boar Hunt, Bear and Swan Hunt, Falconry and Fishing, Roedeer and Heron Hunt* were probably ordered by King René of Anjou, who inherited the Apocalypse tapestries, for his daughter, Marguerite of Anjou at her marriage to Henry VI of England and those four tapestries, made about 1440, went to England and are today in the Victoria and Albert Museum. The tapestries were made probably in Tournai, they are of early Tournai style, influenced by Arras, or else, were made in Arras, or in Walois, the workshop that specialized in hunts. It is difficult to tell for sure the source of origin of quite a number of tapestries of this kind. In 1370 a book on hunting, *Livre de Chasse* was published in Paris, written by Gaston Phoetus, count of Foix. The tapestries are based on this book. The large compositions (4.09 meter x 9.90) are full of figures, elegantly dressed, or simple hunters, spearsmen, the scene — a hunter's dream — is teaming with wild beasts, fowl, dogs. The horizon is high up with trees, castles in the background.

Later Tournai tapestries are even more complex, crowded, still they keep the two-dimensional character. The sky is reduced to a band on top, figures are in a dense composition, several episodes of the subject are shown simultaneously, interwoven. The earliest document on Tournai workshops is the "Reglement of 1398." The weaving mark of Tournai, the tower, was used after 1544. The Bruges mark was an arm with a crowned B.

The tapestry of the *Passion* was woven probably in Arras 1420-30, and donated to the Cathedral of Saragossa by Dalmacie de Mur in 1456. Rocks, and trees divide groups of people and different episodes of the crucifixtion.

Beauvais. A cock flutters its wings, cries out three times on the church clock of Beauvais cathedral. At the ringing of the hour Christ appears. Angels blow their trumpets; at this sign the people representing the population of the earth disappear from the windows and flames create the image of destruction. Angels with the instruments of Passion, the Prophets face Christ. The scales are held by St. Michael. The Virgin Mary, St. Joseph are present. A young girl, Virtue, is blessed by Christ, an angel with a mandolin carries her to Heaven. From the opposite side arrives Vice with the serpent. Christ rejects him after the scale becomes unbalanced. A devil takes Vice to Hell with thunder, music and the ringing of bells. The Last Judgement is over, the windows close on the famous large clock of Beauvais. But this is not all, the clock also shows the planetary system: Mercury, Venus, Earth, Mars, Jupiter, Saturn turn in their proper relationship. 4,000 stars, 86 constellations each with its sign, visible or invisible according to the hour of the day, lunar and solar eclipses, the exact time of nine large cities, and the calendar are indicated. Each day the minute of sun or moon rise and set, seasons, the zodiac also appear on the clock. In fine weather a little boat floats calmly, in a storm it is tossed on the waves. (Made by Verité, 1868)

Beauvais cathedral built in a ground sacred since history fades into legend, Roman temple, Carolingan early church built 997-98, Cathedral of St. Peter started 1247, forever unfinished; choir with radiating chapels; transept of late thirteenth century; highest, several times collapsed nave never completed; continued in flamboyant style, facades from 1500, 1548. Stained glass windows from thirteenth to twentieth century. And in the Cathedral there are the tapestries.

Life of St. Peter, woven in Tournai, cartoons by Henri de Beaumentiel after sketch of Robert Campin, series of ten large tapestries in silk. The Bishop of Beauvais, Pierre de Hellande 1444-1461, presented the series to the cathedral. Six are still in Beauvais, one is in the Cluny Museum, one in the Boston Museum of Fine Arts, fragments in the National Gallery, Washington. The word PAX is seen woven several times on each tapestry; the Hundred Years War was ended. On the tapestry in the Boston Museum St. Peter is decapitated, praying, on his knees, his head, that was cut off, is neatly placed on the grass next to him, among pansies, dasies, of the "milles fleurs."

There were small tapestry workshops in Beauvais working for local nobelmen when the King, Henry IV became a patron; he gave a patent — an exclusive right to produce tapestries in Beauvais-Picardie, to Louis Hinart (1664-1684). Hinart was a native of Beauvais, weaver and marchand, he had workshops in Flanders, sold tapestries to Paris. He received 30,000 livres from the King for building the Beauvais workshops, and 30,000 loan for wool, dyes, plus he was promised 20 livres for each Flemish weaver he could lure to Beauvais. He started with one hundred weavers, in six years he had six hundred, with fifty apprentices. The King paid him 30 livres per year for the keep of each apprentice. Apprentices had to learn for six years, then work in Beauvais for another two to become master weavers. Foreign workers were recognized and treated as citizens after eight years of work without naturalization papers. The buildings grew into a complex of workshops, for painters, dyers, beer-makers, the baker, of course; all of these shared the privileges, tax exemptions of the weaving workshops. The master weaver had a right to sell, transport tapestries, and bring in wool without duty.

There is a commemorative tablet: "King Louis XIV relaxed this shadow, 1686". Philip Behagle was the head of the workshop; they were weaving the *Victories of the King;* one of the series is the *Battle of Landscrona.* The King, in superb costume is pointing to the conquered city (he had so many places to point to on these tapestries). Large figures in the front are recognizable portraits, behind is the battlefield in perspective, chariots, more soldiers all in an elaborate bordure of arms and armour, trophies. The Beauvais workshop signature was a B and the French Fleur-de-lys.

Beauvais was also weaving the *Acts of the Apostles, Chinoiseries* and from the start the popular *Jeux d'Enfants,* children-putty in verdure, landscape with birds, fountains, flowers. Prices were according to finesse of weaving, silk and gold thread per meter: 45 livres, 59, 62, 90 and up to 100 livres. The children, putty, were tremendously popular and the series was repeated several times. The fat little children carry heavy flower garlands, tools, water the garden.

It is interesting to follow the yearly production: among others, 1723 *Children playing* after Damoiselet, series of six pieces; 1724 *Verdure with children, Chinoiserie, Grotesque;* 1725 *Verdure with birds,* Tenniers, with border; 1730 *Amusements Champêtres* by Oudry; 1732 *Comedies Molière* by Oudry — Molière's father was a weaver; 1734 *Metamorphosis of Ovid* by Oudry; 1741 *History of Psyche* by Boucher; 1749 *Love of Gods* by Boucher — the tapestry of *Mars and Venus* is a picturesque, airy composition, putty flies in the clouds, there is a garden scene; Venus is reclining on a sofa, a softly upholstered gold-leaf piece of furniture, outdoors, with draperies. There is an urn, a fountain and the bordure is like a gilt frame.

Paintings by the impressionists were woven, Cezanne, Manet's *Nympheas* copied in 2,000 nuances, instead of the decorative Gauguin. In 1926 the workshops were weaving a feuteuil by Dufy.

Aubusson And Limoges. There is no proof but I like to believe the story that the founders of Aubusson tapestry workshops were the Saracens retreating after Charles Martel's victory at Poitiers. The sheep were grazing in the pastures along the Creuse — then as today, the fast running river was there to wash the wool — and the lovely girls . . . the Saracens settled in Aubusson and started the weaving . . . the magic carpets, the tapestries that carry us away into a dream world . . . *the Lady of the Unicorn* woven in Aubusson . . . verdures . . . Chinoiseries . . . Lurcat . . . Domjan.

There is reason to believe that Marie de Hainaut, Countess de la Marche and related to the Countess of Flanders, brought weavers from Flanders to Aubusson and Feletin in 1327-1342. Both in Flanders and in Aubusson the weavers' patron saint is Santa Barbara. Weaving methods — the low warp loom are similar.

Henry IV gave his support to Aubusson, Colbert raised it to Chartered Royal Manufacture in 1665. King Louis XV sent experts from Paris "assortisseurs" in 1773 to visit the factories, inspect and supervise the production and instruct the weavers on the selection of colors, nuances in dyeing the wool, and the sets of colors to be used for faces, figures, animals, water, sky, fruit and flowers. Aubusson mark was MRD, MRDA, MRDB (Manufacture Royal d'A).

Not far from Aubusson, Limoge, the city of enameling and porcelain, had produced tapestries. In 1640 Gilbert Roguet of Aubusson, master weaver, married to Jeanne Boffinet, also of Aubusson, established his workshop in Limoges in the parish of St. Maurice. His sponsor was the Bishop of Limoges who became godfather of his son, born in 1642. The son also became a weaver in Limoges. We know of several workshops active in Limoges during the seventeenth and eighteenth centuries. Other Aubusson weavers went farther away to work, for instance to Germany.

Today Limoge is a thriving modern city, capital of Limousine, famous for porcelain and other industries while Aubusson and Felletin are isolated tiny industrial villages of tapestry tucked away between hills of the Central Massive. Functional and unattractive weaving workshops follow the capricious line of the Creuse river, the narrow streets, steps up and down. There are no luxury hotels in Aubusson nor sightseeing tours. Between the old cemetery on one end of town and the old unadorned hilltop church on the other, inhabitans of Aubusson are industrious, economizing and highly specialized in skills to produce their one crop: tapestries. Limoge is their faraway metropolis, Paris is for many an unattainable dream, and the rest of the world doesn't even exist. On crumbling grey walls everywhere, out of place, gaudy posters offer Wild West movies played by the Main Street movie house. Weekdays mean work, happily or at least willingly done and Sundays mean togetherness with the family, sitting with a glass of wine and complaining about politics, rising of costs and unmarried daughters. Eight year old Pierre proudly announces he will become a weaver like his father and grandfather, he already has his portable little loom and teaches his schoolmates how to weave. The weaving workshop is a family business, wives helping with office work, and during the short tourist season in the shop. The small, framed commercial pieces are still here, the cat and the ink-pot for the vacationing "concierge" with modern pieces and rewoven old fragments. Many a fine masterpiece of tapestry from Aubusson looms left Aubosson and can be seen in museums, in public buildings, private collections, in galleries in Paris and elsewhere. Not every weaver's son became a weaver; one of them became Molière, who gives this advice in his play; "Sganarelle ou le Cocu Imaginaire" — when it comes to cheer up Sganarelle's daughter, Celie: "Mois, si j'etait à votre place, j'acheterait une belle tenture de tapisserie de verdure ou a personnages, que je ferait metre dans sa chambre pour lui rejouir l'esprit et la vue."

I, if I was in your place, I would buy her a beautiful tapestry of verdures — or of figures, which I would put in her room to cheer up her spirits and the view.

The tapestry, however, has a more important role in our lives; it belongs to the man in the street; to everyone. In Paris, in France, a great number of tapestries are to be found in public buildings, from the Latin school of little girls to the Eiffel tower and the boat La France, tapestries decorate the walls. The UNESCO, Maison de la Radio, Ministeriums, Universities, Perfecture de Police, Supreme Court, Conservatoire of Music, of Drama, Chamber of Commerce; and in the little towns in the Town Hall, theaters, airport, banks, libraries, business and company buildings have tapestries. Many were specially ordered for one specific wall — not only the size, but thematically the subject was choosen for a specific place; an institute of meterology, agriculture, aeronatics, pharmacy — the artist made the cartoons with this in mind. Not thematic 19th century ordered work but free association, symbols.

How very difficult it is for a small town to establish its identity, how very often it is unwanted; a crime, a racial riot that is remembered; how very different an identity if a small town is remembered by an important art work at a public place.

If there is a wall, there is room for a tapestry; and the place immediately has an image. There are no installation costs, no maintenance, no depreciation of values, the value will only increase — a tapestry is everlasting.

Color in Tapestry

Aubusson   Dongen

The Master Dyer. "It is the color that makes the tapestry," said the master dyer of Aubusson.

The dyeing of fabric and yarn has been most important since the earliest days. The Britons, during Roman times, were dyeing cloth in shades of blue; a vegetable dye, woad leaves, were used. The Egyptians used salts and acids for dyeing. The Phoenicians were excellent as dyers in the ancient world. They got a purple dye, in red-blue shades, from a mollusk in the sea.

During the Middle Ages most colors were obtained from vegetation growing in the countryside or in the dyer's backyard. There were fundamental dyes. Indigo produced blue, weld produced a brilliant yellow (that faded in time), madder produced several shades from red to brown. Sepia was derived from iron oxide. An insect called "Kermes", collected in France, gave the purple color. Out of these few fundamental dyes lighter and darker shades could be produced and their variations, or mixture provided the medieval tapestry maker with all the colors that were needed.

For the background the master dyer made a batch of blue; when the weavers ran out of it more had to be dyed, but the color was never exactly the same shade. This difference, a little paler, a little stronger, gave color a variation, liveliness, change — a hand-made quality and charm to early tapestries.

The medieval eye called for bright, clear local colors and this was right for tapestries. The magnificent series of large tapestries like the *Lady of the Unicorn, the Angers Apocalypse,* were woven with this limited number of colors.

When Aubusson became part of the Royal Tapestry workshops, a master dyer was sent from Paris to prepare the wool for all the looms in Aubusson. During the seventeenth century in Bruxcelles, Gasper Leymiers was master dyer, coloring the yarn for the whole tapestry production of Bruxcelles.

We have visited the master dyer of Aubusson in his workshop on the river Creuse. He prepares the wool for all the looms in Aubusson. In his backyard the wool is drying on strings — red, blue, all shades, among tomato plants and marigolds. Inside, the workshop is dark, obscure from clouds of steam. Dyes are boiled in large tanks. The wool is submerged in the dyestuff, stirred and pulled out from time to time and compared to the color sample. The master dyer is an extraordinary man; a stocky, strong worker, yet of infinite sensibility to colors. A color on a piece of paper or the color in steaming vats, the color of the wool are different things. There is no formula, the master dyer's knowledge and experience and the patient labor produce the right shade. He has a fantastic memory for colors and knows the color-range of each artist by heart. He is strongly opposed to tapestries in black and white or shades of grey, and is depressed if he has to dye wool in grey and black only. The tapestry is color — color that "sings".

While Les Gobelin was still producing the masterpieces of tapestries by Boucher, tapestry production was already on the decline. One of his directors, Michel Chevreul was a chemist. With a scientific method he composed a palette of ten circles of primary colors, circle chromatique. A circle was divided into 72 scales, each with 20 tones — arriving at a total of 14,000 tones. Chevreul had hoped to bring order to the chaos, but soon there were 36,000 tones and with the usage of mixing two threads of different tones to create a third, there

Mermaid

were almost 200 million possible combinations. There was also a great waste because of the necessity of dyeing 1,456 pounds of wool to get the right shade for a 44 pound tapestry.

The weaving became finer and finer and the production slowed down. Skilled weavers could not produce a tenth of what weavers 300 years earlier used to make.

In Aubusson old cartoons were repeated, mutilated — the inspiration, excitement had flown away as a bird escapes from captivity. The expensive dyeing and slow work resulted in a downfall and bankruptcy.

During and after the second World War a small group of modern artists headed by Jean Lurcat, artist-cartoonier, tapestry designer, innovator, and inspired by early Gothic tapestry brought changes and reform, the result of which was the emergency of contemporary tapestry.

We are still to witness a Renaissance at its beginning. There were 150,000 tapestry weavers in France and Flanders during the fifteenth century. There were less than one hundred active before the World War. Today there are 200 working on contemporary tapestry in Aubusson, 40 at Les Gobelins — even less in Beauvais.

In 1930 Mme. Cuttoli commissioned tapestries to be woven from Braque, Picasso, Matisse. When these artists created their paintings they had no thought of tapestries, nor did they have any relationship with the weavers or control over the work. The tapestries were exhibited framed. It was not the real thing, but a first step.

Experimental Tapestry Today. Lurcat was a great innovator, but certainly could not foresee the development of experimental weaving, art with yarn produced in the last couple of years. He was experimenting but always within the framework of traditional tapestry; that is, the square format, flat surface and even tapestry weave. In Paris, in Aubusson the traditional weaving prevails while a lot of experimental work is done and innovations come from elsewhere.

In the United States of America hand-weaving enjoys a new popularity and there is a great revolution in the art world, art made out of yarn. Fresh new art forms for which we did not even find the right name are harbingers of a new epoch, along with the renaissance of the crown of weaving techniques: tapestry.

There are works called fiber constructions rather than tapestries, free standing sculptures made of yarn; stuffed figures of textile; soft or solid forms; ribs covered with fabric; constructions of metal or wood supports are covered with yarn, knotted or woven without a loom. Woven or knotted structures, like gigantic lampshades hang from the ceiling; tent-like shapes stand; there are mobiles, and kinetic pieces depending on body movement. Hangings are woven, with open areas and incorporating pile-carpet knotting, or other mixed media. The techniques used include every possible variety and combination of weaving, netting, basketry, crochet, knitting, applique and embroidery. The artists use the whole repertoire of old techniques, they even borrow ideas from bird-nests, wasp-nests and with great virtuosity create their own. The materials used vary from shocks of raw flax or bunches of unspun wool (impossible to resist the temptation to stroke, pluck) — to plastic fibers, raffia, reed, sisal hemp, cow hair, all woolens and telephone wires. Glass, ceramic, metal objects are often incorporated. The statement of each piece grows out of the technique itself.

One of the revived techniques is *macrame,* the word itself and the art comes from the Saracens, then from Southern Italy where it became an intricate lace used in church art during the 16th century. It spread North and was the sailors' hobby in Queen Victoria's time. It uses string, yarn and rope; these are made into two dimensional or three dimensional works by the use of various knots. The encyclopedia of knots contains 3,400 different ones. There are traditionalists and other artists using macrame in a more individual and avant garde way.

Tapestry, traditional or experimental, the innovations in fiber are all fine arts, new original creations of the artists' unique, often one of a kind, or else like bronzes can be repeated in a small edition. In the art world the last decade saw the complete denial of craftsmanship and such movements as non-art. Revitalized, modern tapestry has once more reoccupied its position of leadership. Experimental art with yarn lit up a new fire of forceful creative powers, exciting, stimulating, — and brought back unexpected as if by magic what was feared to be lost forever: craftsmanship.

This new art with yarn still needs to be evaluated and its impact on future art movements cannot be measured today.

Experimental work is often no longer interested in color and uses shades of natural fibers, or all white or black and takes the place of sculpture while pictorial qualities survive in traditional tapestry. There are artists who create works for one exhibition to be dismantled afterwards, some use materials that disintegrate, and place their work out in nature, in parks to become part of nature, mellow with rains, like dead branches, get thrown in the wind like cobwebs. Other works are made for eternity, like tapestries. The variations are endless and there is no limit to the artists' creative power and force of imagery.

In Aubusson in 1945 the involvement of modern artists in the weaving workshops brought about the modernization of the workshop, simplification of technique. Generally the aesthetic tastes have changed with the times and in this the artists were forerunners. Instead of the perfectly finished lifelike tapestries, the crude, early ones were creating an echo. The rediscovery of Medieval tapestry, the return to robust, healthy earlier weaving with a limited palette made the rebirth of tapestry in our time possible. The over-refined weaving, the tens of thousands of shades, the silk, the gold and silver thread were all discarded. The pure terrestial, peasant wool became the only material used.

Lurcat, with the master weaver Tabard, worked out a shorthand method of numbered cartoons. In this case the cartoon is not painted, only the outlines are marked and instead of colors there are numbers. The artist and the weaver agree which red to use in six shades. Extremely conscious of the time element involved and the costs, this easy to read method worked perfectly for Lurcat. Certainly an abstract way of thinking is necessary for an artist to think of number 50 instead of blue.

The Domjan tapestry has a limited palette, a perfect freedom of picturesque expression is achieved within the limits set by the artist alone.

The Fire Peacock, Tapestry by *Domjan*. Pale carmine rose tulip-shaped wings encircle its swan neck and small head. Like gold wrought treasures of the animal style, animal shapes abound; a fish in his eye, a snake on his neck, and birds — at least twenty-five of them — on his body — a puzzle, try to count them. One gets a different count as the room gets darker at dusk or from another part of the room. *The Fire Peacock* is one of Domjan's "milles oiseaux" tapestries. The birds in positive and negative shapes broken up into a tour-de-force jumble of details; geometric ornaments wave, triangles, dots, parallel and crossing lines, tulips, heart-shaped. One bird has a string of pearls in its beak — and among the motifs is the weaver's comb. Motifs are interwoven — one flows into another, inextricable.

The background is of several shades of dark purple. There is no horizon, no sky and no earth in the land where these magic birds came from. Their three to four wings do not correspond to known bird skeletons, the feet end in leaf form; yet all is truer than reality by the artist's force of creation. *The Fire Peacock* is a being beyond its radiantly beautiful appearance; it creates an atmosphere.

The colors suggest fire, sundown, an afterglow when skies turn purple, clouds glow from the reflection — a swing, a sound of wings, the bird may fly away never to be seen again — and it will be night.

Colors — let us take red; looking around in the studio right now, we see several tapestries in which red colors dominate; *The Fire Peacock*, the *Red Cock, Bird and Young, Paon,* and a knotted rug, the *Black Poppy*. Had Domjan used the numbered cartoon system, one or several pre-dyed red colors in 4-6 shades, the reds would repeat in the different tapestries. Instead each is another unique red world by itself. It is very difficult to describe colors, but in these tapestries the artist's fantasy run the gamut from empurpled rubies and orange tinted reds to pinks to violet red and antique amber. Not only is the basic dyed red different each time, but it is set with other colors and by contrast or similarity a different effect is created. *The Fire Peacock* is a symphony of delicate pale shades into dark; the *Red Cock* is more robust, its red-yellow combination almost provocative; *Bird and Young* is vibrating by the clashing and harmonious cold and warm orange and rose; *Paon* is built on complementary colors of red and green, the background is the red; the *Black Poppy* is the unusual combination of red and black on a dark olive background.

Paon wears three elegant peacock feathers on his head; he is looking backward toward his leafy tail. The shape of the bird is well balanced to the background area; the straight, long neck penetrates into the background, its cold, steel-blue, green stands out on the deep warm red which in turn cuts into the bird shape. A simple leafy motif decorates the body and is continued in silhouette outside of it. The steel-blue becomes leaf-green and lights up at the end. The color combination defies color theories which state the red comes forward in a red-green composition. The well-defined shape, clear simple design make the *Paon* a perfect example of tapestry art.

The *Red Cock* strong, powerful, fills up its space and hardly any background is left, bursting, sparkling star elements invade what little is left of it. The stars, each different, are lighter in the center, orange and cadmium yellow on brick — to rose. The stars are lacy, through small lines and shapes the dark blue of the background shines through; this lace effect breaks up the monumental shape of the cock into latticed arabesque. The cock, symbol of the sun, appears here radiant, victorious over the night sky.

Bird and Young. Tapestry by Domjan; the long, straight neck of the peacock bird contrasts with the circular shape of its tail-plumage. The bird, with a tulip blooming on its head, is talking anxiously to its numerous offspring — he is calling them; two of the small ones are hidden in the ornamental foliage of its tail, they are hard to find. One ducks under the mother-bird and another one is perched atop. Several others — in a light rose linear design — populate the dark brown shade of the background. Each little bird is a personality by itself, different from the others in shape and ornamentation — did the peacock-mother hatch a basket of painted Easter eggs?

The topmost bird is a funny character; a three-leafed clover grows from its red wing. Its breast is of a purple-rose checkered pattern. The bird below is more like a peacock; it has a three-fold, fan-shaped tail. This young peacock holds a leafy branch in its back on which it stands. The birds inside the large shape of the mother-peacock and predominantly cadmium ocher, and pink, interwoven in the many hues of rose to orange to pale lilac of the bird with its well-defined shape against the dark neutral background. BIRD AND YOUNG has a simple decorative form and a great variation of charming details; a lovely warm, and humorous tapestry — for the boudoir of a modern princess.

Knotted rugs. In case of knotted rugs — tapis à poil ras — haute lisse, high warp looms are used; wool knots are made, later cut to make the pile; the second row is of linen thread or wool that makes the canvas seen from the back of the rug or pile carpet. There are different knots like Turkish, Persian. Persian carpets conquered the world from Venice to Barcellona and Paris. They were imported and carpets were also produced in Europe in that method — tapis de Turquie. In 1604 Pierre Dupont and Simon Loudet working in the Louvre workshops received the privileges to make rugs in the Oriental manner, with gold and silver thread. In 1627 the workshop was transfered to Chaillot in an old soap factory — from here originates the name savonerie carpets (savon-soap). On haute lisse looms wool was used and linen for the second row, the canvas of the rugs. Louis XIV gave them the title, "Manufacture Royal des Tapis facon de Perse et du Levant" — later united with Les Gobelins.

Rugs, and pile carpets were a national industry of great importance in Persia, Turkey and the Caucasus. The first well preserved piece comes from 500 B.C. During the 17th century and later they reached their highest level of perfection. Carpets were small prayer rugs carried by the Mohammedans with them to immense royal size rugs for covering palace floors. They were made of wool with cotton warps and specially rich carpets used silk instead of wool produced in Persia. Caucasian rugs, like the nomad carpets and tapestries, were non representational, of severely abstract geometric design. Persian rugs were stylized but basically representational, older Persian rugs include several types of design; the medallion, vase, garden, floral and animal carpets. The makers of carpets are not known by name, they were the patient, dedicated, fatalistic artists and artisans who spent their lives to produce beautiful works that survived them. Rugs appear as part of the design in tapestries like in the *Lady of the Unicorn.*

Traditional Pattern

The Black Poppy, knotted rug by Domjan. In a monumental design the artist combined a large red and orange open petalled poppy with a half-closed black one. Stems, leaves, and the closed bud are of bluish green as well as the seed pod with delicate black pistils. The olive green background in several darker and lighter shades helps the red color to come forward, bold, and bright. Folds on the petals are drawn in strong lines, red on yellow, and yellow on red — and dark grey and black for the black flower. Knotted rugs do not permit the delicacy of woven tapestries.

Among tapestries in blue colors *Peacock at Night* and *The Proud Peacock* can be compared, both are large, upright tapestries, each with another-worldly bird in blue filling the composition. They came from the same period and have some similarity in the elaborate details but here the parallel ends.

Peacock at Night has a small delicate body, tiny head and an immense irregular enclosed shape for a tail. Starting from the small head peacock feathers form a crown, branch out, and fill the space of the tail, crossing, inter-crossing and parallel feathers bend, curl, the eye of these peacock feathers is the scene of many interesting happenings; animal shapes, stars, circulate in its own universe at its tail, lines of delicate peacock feathers behind them a heavy purple shadow, curl back toward the large tail shape. The graceful and elegant bird composed of many fine shades of aqua, ultramarine, sapphire — green on the lower wing — has highlights in red and yellow; the red eye, edge of a tail, on the wing, there are a great many different colors on this blue peacock on the dark blue background, and a great diversity of elements within the monumental simple outline.

The Proud Peacock doesn't have a big solid mass of shape; instead, the outline leaves more space for the background; positive and negative areas interplay; wings and feathers are rather ribbons that flutter in the wind, the space between them is as important as the ribbons themselves. There is a great unity and a sparingly limited palette of shades of light turquoise, aquamarine on darker green shades, on a darkest green background that in turn has pale blue ornamentation. Variations of small peacocks and flower motifs fill the wings and body of the bird and continue in the background. The peacock's small crowned head rests on the pointed neck, the haughty bird is looking upward. A plump body and small feet give a touch of humour — the slender graceful swing of the lines a touch of elegance. The full body fills the lower part of the tapestry while the upper part has a floating lightness — the bird it-self expresses perfectly the crystalized essence of birdiness as if it was a crystal-ized form of cloud and sky, in its blue — an eternal yearning for distance is herewith fulfilled.

We may recall the robust *Blue Cock,* the pale blue-grey on deep dark blue of the *Devilish Deer,* the blue-gold combinations of *Peacock of Carnations* or *Moonshine Peacock* to get the feeling of the blue world of Domjan.

Savages have a ball

Savage Lady

Savage Tapestries. Edward Manet's *Dejeuner sur l'Herbe* arroused a scandal at the "Salon Des Refuses" in 1863 not because of the nude but because of fully dressed gentlemen around the nude, yet . . . Elegantly dressed personages dance with savages in a large, complex composition of figures in a landscape on a tapestry in the Church Notre Dame de Nantilly in Saumur. The tapestry, from the 15th century, is one of a great many that have savages; men and women bare-footed, dressed in nothing but tufts of animal-like hair. Under a canopy musicians are blowing their wind instruments. In one corner elegantly dressed persons step out of a tent. The horizon is high up, there is a forest behind, with monkeys (for symbolic reasons?). A savage lady in a tall and ellaborate headdress and a rich cloak that barely covers her, gives her hand to a fully dressed aristocratic youth. Savage couples, and fashionably dressed couples, in silken brocades and velvets, are dancing. There are knights in armour, pages — and nobody seems to be astonished at the undressed figures among them. In the foreground and between, where little space is left grow the bunches of flowers of the "milles fleurs".

The French court staged "Balets des Sauvages" and we know of one instance when the King, Charles VI appeared at a ball "Bal des Ardents" in the Hotel Saint Pol, Paris, in 1393, in savage costume. He and his entourage had waxed their bodies and rolled into unspun wool. Fire broke out at the ball, the wax melted in the heat, few of the savages were lost in the fire. The King's life was saved by a lady-in-waiting who wrapped her cloak around the King and led him out of the ballroom.

An early example of savage tapestry, called "Wildemanner Teppich" and also "Waldmanner Teppich" in German (wildmen and forest-men) comes from the Strasburg Bishop's Palace in Karnten and is now in the Klagenfurt Museum, Austria. Savage girls with flower and leaf wreaths on their blonde locks and wild men are holding imaginative animals on a leash; the animals are two-footed and quadruplets, combination bird-camel-griffin-unicorns. They are being beaten with iron clubs by the wildmen whose bodies are covered with red-blue undulating hair. The background is dark green. Scriptures about true and false love — with some undeciphered letters, perhaps initials of weavers — complete the scene and indicate a connection between the wildmen representations and romantic love and sex.

Seven persons, three men and four ladies are participating at a hunt with falcons on a tapestry in the Kunsthistorisches Museum, Basel, made for the tax-collector, Peter Fehr of Luzern and his wife, Benedikt von Meggen; they were married in 1491 and his coat of arms is woven into the tapestry. Fashionably dressed figures are intermingled with savages in savage outfits; barefooted and hair-covered. The figures stroll in a scroll-foliage interwoven with dogs, birds, coat of arms on a dark background. Around each figure there is a ribbon with lettering, no need for a guide or explanatory catalogue for this kind of tapestry — the Middle Age weaver has put the story right into his composition. A young man's falcon throws himself on a rabbit, he is addressing a wild woman and says; "Fair Lady, don't be afraid, the rabbit can't bring bad luck." The wild lady doesn't seem to be reassured and says on the ribbon of text woven around her; "My good hunter, save me and watch the dog that he doesn't bite me."

The origin of the savages is a mystery — where did they come from, who are these savages who conquered the fantasy of fifteenth century society from France to Austria and Germany, in literature, plays and tapestries? Does the idea originate in antique pagan mythology, satyrs, Silenus, salvi homines? . . . Or else were the savages, the wild men a reaction against the excesses of fashion that reached the end of the line with the first fashion shows? Wood-doll models circulated from Paris, dressed with utmost luxury in furs, silken brocades. Headdresses of the time were so high they could not be built higher. Pointed shoes were so long, the point had to be chained to the knee to keep the wearer from falling. Were the savages the Middle Age beatniks — a protest?

The forest — much wilder than any we know today, and mostly undiscovered was populated with imaginary animals — unicorns — as much as real ones and still had the character of the fabulous — where anything can happen. In the forest, in perfect freedom the savages lived their unbound life and love far from courtly etiquette and duties of the knight.

Like the Rococo's longing for simplicity was expressed in the shepherd's plays — so the fifteenth century turned to the savages for excitement and amusement.

In the Furstliche Galerie, Sigmaingen is the most beautiful savage-tapestry, from the fifteenth century. A horde of wild men are hunting, fighting against lions, dragon, unicorn. One scene shows a savage woman nursing her child — this scene is very much like a Madonna and Christchild with St. John and the Three Kings; here are the three savages bringing presents, one of them riding a deer; is this a joke, it almost seems a sacrilege — or else the savages represent a pure life, a back-to-nature, Gothic Rousseau's ideas. Characteristic to German tapestries of the period, the foreground has regular, repeated lumps, little hills with grass and flowers or a rabbit — kind of primitive "milles fleur". Oaks, linden trees, pines, treated like textile ornaments fill the background. The contour lines are black, there are strong local colors in three shades, a limited palette and no hatching.

In the Regensburg Savage tapestry decorating the Rathoussahl, family scene of savages, a deer hunt and the seige of the Miennenburg-castle are shown above the lumps and between ornaments of oaks, birds, branches. A wild girl is riding on the back of a wild man on his fours, (a Middle Ages game) this scene, the story of Aristotle and Phyllis, was used frequently as a comic representation.

A smaller tapestry comes from Nurenberg as can be seen from heraldic symbols and costumes; on it a group of savage men and women with pitchforks on wild goats are riding out of a walled-in city. Two men are tied by a girl. Forests, buildings, and woven ribbons of text complete the scene.

Savages are mounted on fantastic animals, are seen in love-scenes, and hunting in front of an Oriental background of pomegranate motifs, and birds, in green with red, on a dossal from a Basel workshop from the mid-fifteenth century.

Savages are partaking of a banquet in a tent, fantastic animals attack a castle on a tapestry from Prague, now in Vienna.

We know the artist-designer of one of the tapestries; Stecher, Meister der Spielkarten designed the Wildmanner tapestry made in a Basel workshop in

1460, now in Kunstgewerbe Museum, Vienna; wild men in occupations like the calendar pictures; sowing seed, ploughing, hay-making, mowing, harvesting. The figures do not fill the surface but are arranged loosely in front of an ornamental background of oak leaves. The branch-woven fence is drawn from life. Swinging ribbons with Gothic lettering explain the scenes.

Red and blue haired wild men ride the dragon-like chimara between oak trees and thistles, symbolic erotic scenes are on a tapestry made in the Frauen Kloster im Bruch bei Luzern, now in Zuricher Landesmuseum.

Recently discovered frescoes from the Middle Ages in a rich burgher's home in Basel have elements similar to the tapestries, live-scenes in a garden where little savages play.

A tapestry In Basel , about 1477, illustrates the medieval chanson Von dem Grafen von Savoi. The count, improverished, has to sell his wife to pirates. The countess is taken to France on a sailboat. The King wants to marry the countess. She begs him to wait for a year. The count wins over the king in a tournament and finds his wife. The king reunites the couple and returns their lost properties. This was a favorite song of the Miennesinger. The strangeness in this tapestry is that except for the tournament where the two heroes are in armour — all participants, count, and countess, the king sitting on a triangle-decorated throne, (as if designed by Vasarelly) are all in savage costume. The tapestry has an oak-branch background and no horizon, the story is told in Gothic lettering on ribbons, in Besancon Musee des Beaux Arts. It was suggested by certain experts that the savage tapestries are illustrating a piece of Middle Ages literature, now lost; this tapestry certainly proves the opposite; the savages subject was much more wide-spread as to trace it to one novel or song. The characters in the tapestry are in savage outfit because it is a love-story.

In the Mainz Dome treasury is the late Gothic savage tapestry that shows the influence of woodcuts of the period. Thin, elongated Giacometti-like figures, wild and otherwise, are out on the hunt with spear and dogs. Trees and plant motifs in the background still recall the Byzantine tradition, that lived that long. New is the use of a yellow color in the tapestry.

Although traditions of the Middle Ages were continued in tapestry well into the sixteenth century, few are the topics that were on the scene that long. A Wappentpich (coat of arms, heraldic tapestry) from 1621 in the Schloss zu Sondershausen has savages and angels woven with heraldic motifs.

The Last of The Savages. Savages, wild men are still around as folk masquerades. Yearly festivities can be traced back to ancient Roman, Byzantine religious rites, or older roots; Vogel-Gruff-Tag, in Basel, claims to have Egyptian origin The days of celebration are either the beginning of a season or the end of a past one, to keep the evil spirits away. The meaning of ceremonies faded into obscurity. The unnatural movements: run, dance steps, unusual sounds: chimes, yelling; the prescribed costumes, animal forms, show the mystic supernatural origin of these folk festivals.

Men are dressed as Wild Men, "Tschameler", at the Cowherd's Fair, "Alplerkirbi", at Lake Luzern with Bear, Goat and Horse participating.

Wild Men, Harlequins are having a show at the "Schemenlaufen" in Tirol. In Basel the magic of ancient days is recreated each January at the Vogel-

Gruff-Tag. There are three burgher societies of Klein-Basel active today in their official functions; their coat of arms, heraldic symbols since medieval times; the Wild Man, the Lion and the Griffin are personified and acted on the day of the festival.

Starting in the early morning hours, the four "Ueli" in Harlequin outfits, run in four directions collecting alms for the poor; they will not stop until late evening. Cannon shots announce; the Wild Man has started his trip on a raft on the Rhine, accompanied by two drummers in powdered wigs and historic costumes and two bannerets holding guild banners. While the Wild Man floats downstream, from the other side of town the Lion and the Griffin come to meet him.

The Lion "Leu" is symbolizing strength expressed in special energetic dance steps. He also is the symbol of the sun and the light. His mask is of copper and his brown gown is decorated with tufts of orange and yellow wool.

The Griffin "Vogel-Gruff" (phoenix?) stands out tallest of the three with a long neck, outstretched wings, a gold chain on his neck. He has claws on his feet and the gown is of leather. His movements are slow, but it must be quite an accomplishment to carry the forty pound headdress and bend hundreds of times saluting the Wild Man, city officials, dignitaries and the public all day. Costumes are restored each year, some parts are two or three hundred years old, others have had to be replaced as they have gone through such a strenuous day each year.

The Wild Man has the hardest role. He is prancing and dancing to the rhythm of drums, he is yelling and shouting. On his shoulders he carries a young fir tree, uprooted, with the earth still in the roots. His costume is green. His large mask of an angry face is of copper, decorated with evergreen foliage and little green apples. There are apples also in the branches of the fir tree. Boys try to snatch them and, as part of a game, the Wild Man chases them away. Even so, the apples are soon gone.

The Wild Man is the heraldic symbol of the Hares, one of the three burgher societies; a company of fishermen, hunters, landowners, goldsmiths, lesser nobility and today, of any honest Klein-Basel citizen of good standing pursuing his trade there. On their banner is a yellow hare with a red background. The Wild Man stands for the Damon of fertility, life on earth, germination, growth, rebirth. He is adorned with red apples and ivy leaves.

He plunges his uprooted fir tree in the fountains here and there on his way and sprinkles the people around with water.

When the three masked personages, Wild Man, Lion and Griffin meet at the river bank they greet each other and dance across the bridge between Klein-Basel and Basel (the first bridge was built in 1225); then proceed with drummers, bannerets. There are a few traditional stops. In front of the Town and Cathedral Museum the presiding Guild Master receives the greetings of the three figures, who also honor new leading citizens' houses with their dance. They are invited for food and wine and are received in a large banquet hall. There are songs of wine and Rhine, songs of gratitude, toasts and speeches, then the three personages are back in the narrow streets dancing and marching, ending at night with a procession with lanterns.

The last Savage

German Tapestry. Two Dominican nuns — white gown, black head-dress — are working at a small upright loom; one sits and weaves while the other, standing behind her, separates the threads. A basket of wool is under the chair. This little scene and an initial with a cross serve as signature of the makers of a tapestry, the *Passion;* in the treasury, Bamber Cathedral. This tapestry was probably made in the Dominican women cloister "zum Heiligen Grab" in Bamberg with another tapestry, *The Three Kings* that also has a weaving nun — self-portrait of the weaver.

The history and development of French and German tapestries is very different; while French tapestry served the King and princely patrons, aimed at large size for upmost luxury, and was made by professionals; German tapestry remained a home industry. Even if made for sale, German tapestry was much smaller in size, produced by a modest trade for rich burghers. Most pieces were not intended for hanging on the wall, but used as dossiers, pillows, or curtains to close bed-alcoves. One cannot talk of designing artists, or cartoon makers; there was no large apparatus. A woodcut or frescoe painting was imitated by a person who did not give his life to weaving, but tried his hand at it. Most tapestries were woven in the monasteries for their own use, dossals, hangings were a protection against cold and draft in the unheated churches. The monastery provided protection, here widows and noble-women led a cheerful life; besides the cultivation of arts and letters they were working on weaving and embroidery. The tapestries were narrow stripes, illustrating scenes of the bible, legends, imaginary animals or heraldry, in strong, clear colors, and a coarse weaving.

The earliest known tapestry of the Western world is the *Cloth of St. Gedeon* made in Cologne, early eleventh century. Repeated and regularly placed circles have identical griffin-and-deer motifs imitating a Byzantine silk of the ninth century, but the lion heads at the crossing of vines of the border are typical of the German Romanesque style. Strong line warps with thick wool wefts in red, green, with black contour lines are used. Several other tapestries are known of animals, the sacred tree, in repeated medallions taken from silk brocades but soon an independent new style emerges.

Dorsalien of the choir in Halberstadt Cathedral are figurative; the style is bold, new, decorative, and monumental. The strong coloring is well preserved. The simplicity, naivete, and even the imperfections of the techniques have something of the primitive or folk art, perhaps closer to today's taste than the most perfect pompous tapestries of the last centuries. *The Abraham or Angel* tapestry of the late twelfth century is 1 meter high, 9 to 10 long with large figures, round faced, a round red dot on the cheeks, staring icon-like with large, dark, almond-shaped eyes. The tree-flower-ornament in the background; the caligraphic treatment with parallel dark and white contour lines, zig-zagging drapery folds are from Romanesque manuscripts. There is also a connection with German frescoe painting of the period of the *Abraham or Angel* tapestries as well as the *Apostle* tapestry, also in Halberstadt and the *Charles Magne* tapestry.

Names of the characters, and often the story is told in woven lettering. The *Abraham or Angel* tapestry has a straight border of lettering; The *Apostles* have a straight simple ribbon telling their names. Text ribbons are first func-

Monastic Workshop Domjan

tional, but later become more and more elaborate, with complex curls and bends, swinging around figures as an important part of the composition. On the Halberstadt tapestries the lettering is in the Carolingian alphabet; on later German tapestries Gothic alphabet is used. The alphabet, styling of letters, spelling, misspelling, dialects, plasticity of ribbons, all give a clue as to the time and location of the workshop which made the tapestry, if signature and date is missing. Costumes, coat of arms, heraldic insignia help to identify tapestries.

We do not know if Durer ever made a cartoon but his influence can be seen in the use of perspective, the proportions of figures, in tapestries like the Prodigal Son, from 1517, the shortened square table gives a depth, the figures, dressed in the rich costumes of the period, are more life-like. The weaving is more detailed, haché is used for modeling. In German tapestry the line drawing is prominent in contrast to French tapestry which is always picturesque.

Maximilian von Baylon was a great collector of tapestries; his agents made their purchases from Antwerp to Venice.

Weavers from foreign countries settled in Germany; in Nurenberg and Frankfort, Flemish weavers were working from the mid-sixteenth century. Lucas Cranach's students prepared the cartoon for the tapestry, the portrait of *Charles V,* woven in 1545 by the Bruxcelles-trained weaver Seeger Bombeck, a prolific craftsman, for the Dresden Schloss. The realistic figure leans out of the marble frame, woven; perspective, shadows and a rich bordure make this work similar to other Bruxcelles tapestries. Another highly active master weaver, Wilhelm Pannemaker, had a great number of tapestries woven for the Heczog von Alba during the later half of the sixteenth century. He also made a *History of Alexander* for Philip II of Spain, now in Madrid and a *Romulus* series now in Vienna. In 1604 Jan van der Biest from Bruxcelles founded a tapestry workshop in Munich and started to weave large, formal, historical tapestries. In 1550 Maritz von Sachsen, Flemish weaver started to work in Dresden.

During the late seventeenth century German aristocrats had their agents in Bruxcelles, Paris, and purchased tapestries for their Lust-Schloss, pleasure palace. French tapestries came to Germany continuously as gifts to princes and bishops. The residence of the Freiherr von Geyersche in Koln had a salon of Aubusson tapestries in 1765, Pierre Mercier, "tapissier d'Aubusson dans la Marche" started to weave Le Brun's *History of the King* in Berlin in 1786 with much silver thread in it that later turned dark. In 1699 Jean Barraband started weaving in Dresden, he also came from Aubusson. Charles La Vigne, French weaver, wove the Watteau-like comedia della arte figures in Venetian architecture and park scenes, flower decorated columns for the Charlottenburg Schloss in 1730.

Swiss Tapestry. A sweet mood of spring and youthful magic emanate from the tapestries of the Middle Ages made in Switzerland; young love, a naive belief in the fantastic, with the mysterious woods always in the background. For two hundred years Basel was the center of a flourishing tapestry production. Enea Silvio Piccolomini, the later Pope Pius II and Andrea Gattaro, Ambassador of Venice, write about the tapestries in their letters. In 1451-54 there were three separate home-workshops "Heidensch-Werkerinnen" who paid tax in Basel. Earliest documents called the tapestries "ceuvres paiennes" pagan, which shows their ancient or Oriental, pre-christian origin.

At the Loom

There was a strong Oriental influence by direct commerce. Trade routes from Venice, crossing the Alps, to Basel, to Paris, and to North, brought silken brocades from Alexandria and Byzantinum. The repeat pattern, the geometric design of Oriental fabrics was woven into tapestry backgrounds. The tapestries were made by home-industry or itinerant weavers, in a small size, on narrow, portable looms, in the low warp method; a mirror image of the text was woven mistakenly on the tapestry *"The Duke of Brunswick takes his leave"*, Kunsthistorisches Museum, Basel. Swiss tapestries, like the ones made in Germany, were made with coarse, thick wool; silk was not used, and metal threads only much later. Colors are finer, since the French influence was close.

From 1430 dates the tapestry of young girls and young men in "Zaddeltracht." Wide-sleeved gowns, the sleeves ending in zig-zag, which have chained fantastic animals — symbols of passion and desire; repeated flower and leaf patterns fill the background. There are many tapestries of love symbols. A similar one to the above, dating 1480 has figures more elaborate, in brocaded costumes, the chained animals are more broken up in many details, more fantastic, dragon-headed-peacock-feathered-goat or deer in front of large textile ornaments.

The excellent draughtsman and engraver, E. S. Master, made the cartoons for the *Cardplayers*. Lovers in a tent in the center of the tapestry sit behind a sextagonal table, cards, coins are on the table. Trellis and flowers form the background. In the foreground a rabbit, a snail and the coat of arms of the Judge, Claus Meyer and Barbara Luft. They were married in 1471, the tapestry was woven in 1480, for them.

Traditional topics were repeated even though the style changed; it is interesting to compare two tapestries, both with a loving couple *Liebespaar* — these were made on order and may have been marriage documents, like the Arnolfifni portrait by Jan Van Eyck. The Cloister in Gries bei Bozen made the first in 1440. The two figures in the fashionable "Zaddeltracht" are rather primitive. The foreground is made of earth-lumps; the background is a repeated flower and leaf-motif, a simple scematic flower and a ribbon with Gothic lettering. The second tapestry, woven one hundred years later, in 1548, in Basel, continues the tradition of no horizon, ribbon of lettering, two figures in front of leafy background — cabbage greenery in this case — the rabbit is there in the foreground, always, but the naiveté has been replaced with more anatomical knowledge, the body can be felt under the drapery, figures are well proportioned, rounded, shaded, they stand on their feet.

Buda. King Sigismud of Hungary, German-Roman Emperor, was one of the great art patrons of his age. He built the magnificent Gothic castle-fortress of Buda (destroyed during Turkish occupation) and covered the walls with tapestries. A French traveler staying in Buda in 1432 writes about high warp looms and Arras weavers working for the Emperor, in the workshop in Buda. This workshop was highly active during King Mathias of Hungary and his Queen, Beatrice of Aragon who brought Italian artists and rebuilt the Renaissance palace of Buda. Not all tapestries decorating the royal palace were made in Buda, many came through purchase and gifts. Jean sans Peur gave to the ambassadors of the Emperor Sigismund a large tapestry in Lille in 1416; this was a *Hunt* with falcons with elegantly dressed ladies and gentlemen.

The Raphael Cartoons

Florence

Tapestry in Italy. Giorgio Vasari (1511-74) writes: "Raphael had now attained to such high repute, that Leo X commanded him to commence the painting of the great hall on the upper floor of the Papal Palace . . . The Pope also desired to have ready the Cartoons, which he colored also with his hand, giving them the exact form and size required for the tapestries. These were then despatched to Flanders to be woven, and when the cloths were finished they were sent to Rome. This work was so admirably executed that it awakened astonishment in all who beheld it, as it still continues to do; for the spectator finds it difficult to conceive how it has been found possible to have produced such hair and beards by weaving, or to have given so much softness to the flesh by means of thread, a work which certainly seems rather to have been performed by miracle than by art of man, seeing that we have here animals, buildings, water, and innumerable objects of various kinds, all so well done that they do not look like a mere texture woven in the loom, but like painting executed with the pencil." Another contemporary spectator, Paris di Grassis writes: 'The whole chapel was struck dumb by the sight of these hangings; by universal consent there is nothing more beautiful in this world." . . . *The Acts of the Apostles* series of tapestries designed by Raphael places Italy at a foremost position in the history of tapestries.

Tapestry was reintroduced to Italy relatively late, words in the Italian vocabulary show the Franco-Flemish origin; "arazzi" (Arras work) "panno de razza." The Raphael tapestries in the Vatican are hanging in the Galleria degli Arazzi although they are woven in Bruxcelles, completed in 1519.

Looms were producing fine work from earliest times in Italy. In Sicily where Byzantine and Islamic influences mixed while classical traditions never died out there developed a thriving silk industry. During the 11th - 13th century Sicily was under Norman, Schwabian, then Spanish rule. Under the Norman kings of Sicily there were royal workshops in Palermo. Their speciality was sumptuous gold ribbons, narrow stripes, richly embossed, woven in gold thread, exported to every part of the Christian world to ornament church robes and altar covers.

Lucca was the silk weaving center from the 12th-18th century; 3,000 looms were working by the end of Middle Ages. Florence, Venice also had a thriving silk industry and agents in Byzantinum, Paris, Bruges, Antwerp. Giovanni Arnolfini, representative of a silk house in Lucca was revenue adviser to Philip the Good of Burgundy. He and his Italian bride, Giovanna Cenami, had their wedding portraiture painted by Jan Van Eyck.

During the 15th century velvet weaving was invented in Venice or established after Turkish models and like the silken brocades before now the velvet conquered ladies' hearts and painters' imaginations.

Descriptions of Italian festivities during the 13th, 14th century and inventories has "panni picti". — picture panels mentioned, these may have been tapestries, but we do not know for sure. Definitions of the times are hazy, "panni picti" may have been tapestry, or embroidered, brocaded fabrics, or painted canvas.

In 1389 Valentine Visconti left for France to get married to Philip the Bold of Flanders. In her rich dowry there is no tapestry mentioned, but we know when she arrived she received a tapestry of Arras woven with gold and silver thread and representing Alexander the Great occupying Babylon.

From the 15th century tapestries were lavishly used for every great occasion. The Italians who love festivities and have a great sense for pomp and theatrical arrangements made the most of it. The election of a new pope, the coronation of an emperor, the canonisation of a saint, the triumphal entry of visiting royalties, princely marriages, tournaments were all such occasions. During the Pope's inauguration and the sacred procession from the Vatican to the Lateran, tapestries were hung all along the road.

The Gonzagas of Mantova were great collectors of tapestries. Their castle-fortress had a great many period rooms where tapestries still hung, with dozens of courts, closed-in gardens. This was the largest building complex after the Vatican. During their celebrations the Gonzagas hung the tapestries on balconies, windows, and they were also used as stage decor in the theater in Mantova.

The Gonzagas, in addition to their own collection, often borrowed tapestries from their relatives, the Estes of Ferrara, for baptismals, banquets. Ferrara had its own tapestry workshops, these could not satisfy the demand and cartoons were sent and cartoons given to Bruxcelles, Arras. Tapestries with the HK monogram were made by Hans Karcher in Ferrara, middle sixteenth century.

The tapestries — mobile, animated paintings — floating in the air on triumphal arches, framed with leaves and fruit garlands, next to statues, costumed actors, posing for allegorical "tableux vivants" — serving as baldachin — there was no other ornament that could rival the tapestries in the variety of use and splendor.

Pope Pius II organized a Corpus Christi day in Viterbo in 1462. Each cardinal could choose a certain block on the route of the procession to decorate. Bishops, archbishops, prelates were vieing with each other to produce the most sumptuous decoration. The result was the largest and most magnificent open air tapestry exhibition of history; Arras, Ruen, Mantova's masterpieces. Roderigo Borgia, the future Pope Alexander VI had the greatest success with his "arazzi", this certainly had an influence on his later being elected Pope. Pope Pius II had a large decoration constructed in the St. Francis Cemetery, the gigantic pavilion was like a rainbow of colors — tapestries of allegorical, and historical themes, hunting scenes. According to witnesses it was like paradise.

In 1473 Hercules d'Este was married to the daughter of the King of Naples. For this occassion Nicolas V ordered a tapestry to be woven: *The History of Creation* was woven in Rome and described as the most beautiful tapestry of Christianity.

In 1465 the King of Sicily married his son to the daughter of the Duke of Milan — Pope Paul II lent 6 large tapestries for this occasion; a celebration could not take place without the decoration of tapestries.

The Medici coat of arms has six balls of wool. Generations earlier, before becoming the ruling family of Florence and before being one of the great banking families of Europe, the Medicis made their first fortune in the wool trade. During the 14th century Florence had a population of 90,000. There were 200 workshops of the woolen trades, "arte della lana" and 30,000 men worked in the textile industry. In 1442 the Medici agent, Felice Brancacci, was sent to the Mameluk Sultan in Egypt to arrange direct export of wool from

Florence. On his successful return, Brancacci commissioned Masolino and Masaccio to paint frescoes in the Brancacci Chapel of S. Maria del Carmine in Florence.

The Medicis, great patrons of the arts, were also addicted to tapestries. Piero di Cisimo de Medici owned twenty tapestries in 1456. Tapestries were ordered for the munnicipal building of Florence from Livino Giglii of Bruges. When the work was finished he received, besides his payment, a certificate stating the perfection of the woven figures — "only their voices and breath is missing" . . . "when hung on the facade the building puts on its festive air."

Cosimo de Medici (1519-1574) commissioned Flemish weavers, among them Jan Rost of Bruxcelles, to make tapestries after the cartoons of Bronzino. The weavers, at the height of their art, were well prepared to do tapestries in the style of the Florentine school; with landscapes that were no mere two dimensional verdures, but had atmosphere, light, distance, hazy blue mountains far away, the trees more natural, reflecting a delight in the countryside. The borders were rich with garlands and a vital part of the tapestry. During the reign of the Grand Duke Ferdinand II (1621-1670), Florentine tapestry workshops made copies of Raphael and also used cartoons of Andrea del Sarto. Florentine weaving workshops used the city mark FF on the salvage, or the fleurs de lis between the two Fs.

Vasari was commissioned to redecorate the austere republican townhouse, the Palazzo Vecchio, and change it into the princely residence of Duke Cosmo. Vasari, architect, fresco painter, designer and over-all interior decorator, raised the ceilings, built a wide staircase, used marble, carved and gilded decorations and hung the walls with "arazzo", some of which were woven after his design.

The founder of the tapestry collection of the Vatican was Pope Nicolas V (1447-1445). He called weavers from Flanders, Siena, Florence; Jaquet d'Arras was master weaver; the Pope paid him 200 gold ducats for five pieces he completed. Pope Urban VII (Barberini) enlarged the collection. Scenes from the life of Urban VII were woven. Pietro da Cortona was the favorite painter of the Vatican tapestry workshops. The symbols of the free Papal weaving workshops were — bees, tiara, Romulus and Remus and St. Michael.

Italian workshops had Flemish weavers working on one commission temporarily or settled in Italy, while some of the Italian weavers were trained in Flanders. Flemish merchants had permanent depots in Firenze or other towns, or went from castle to castle selling Arras, Bruxcelles, French tapestry; thus it is difficult to follow the production of workshops or to identify some of the pieces.

The Raphael Cartoons and Tapestries. Raphael's series of tapestries of the *Acts of the Apostles* are the best-known series because of numerous copies in different countries and also because we are fortunate to have the cartoons, some of the most important surviving examples of high Renaissance art.

The tapestries were commissioned by the Pope Leo X in 1515 for the Sistine Chapel, designed to be hung under the frescoes on certain festive occasions. The cartoons were finished by 1516 and immediately sent to Bruxcelles. The master-weaver was Pieter van Edinghen called Pieter van Aelst, who had been "tapissier du Roi" (King of Spain) since 1502. The tapestries were woven on low-warp loom. The cartoons were cut into stripes to fit under the warp threads and the design was woven in reverse. The weaving cost 2,000 ducats for each of the tapestries; a considerable portion of the sum spent went for gold thread and silk. Raphael was paid 1,000 ducats for each cartoon. The work progressed fast and seven tapestries were displayed and greatly admired at their designated place during Christmas, 1519. The set of the ten original tapestries are in the Vatican today.

The cartoons remained in the workshop. Francis I, King of France ordered a set of ten by van Aelst, a set of nine was woven for Henry VIII and is now in Berlin; one set of nine was made for Ercole Gonzaga for the Ducal palace of Montova; Philip II of Spain ordered the nine that are now in Madrid. A set woven in 1620 in Bruxcelles is at Hampton Court. Many more sets were woven in Bruxcelles, without gold, by succeeding master weavers.

King James I of England established the first English tapestry workshops at Mortlake in 1619. Manager at Mortlake, Sir Francis Crane, discovered seven Raphael cartoons four years later in Genova and purchased these. Francis Clen made copies of the cartoons and these were used in the tapestry workshops. The original borders were not woven because of the subject matter — scenes from the life of the pope — and new borders were designed, also very elaborate.

One set with the royal arms of England is in the Garde-Meuble, Paris; one is in Dresden and several remained in England. The Sun King, Louis XIV, owned two sets of English make and wanted to buy the cartoons for Les Gobelins. This, however, was not possible and so the French School of Rome made oil copies from the series in the Vatican. Cartoons were made of these and the tapestry production started at Les Gobelins and the Beauvais Royal workshops as well.

When Sir Christopher Wren rebuilt the palace at Hampton Court in 1699 he built a gallery to display the Raphael cartoons. This was an innovation because only finished works of art were considered worthy of display and admiration. In the workshops the cartoons were handled as aids to the final product; the tapestry. Master craftsmanship was thought to be an inseparable attribute of the art work. Sketches were the artist's private affair and the cartoons a step toward the final goal. How different is our point of view: when even graffiti of the past are collected. Sketches of great masters are greatly appreciated for the first, spontaneous record of creation. The Raphael cartoons are of untold value for the direct brushstrokes of the Master. Prince Albert collected documents on Raphael, studied his work. Only avant garde modern art movements from the turn of this century changed the Raphael cult that at Queen Victoria's time certainly semmed to reign indefinitely. From Hampton Court the cartoons were taken to Buckingham Palace and from there to the Victoria and Albert Museum where they are today.

After the many copies woven from them, the folding, the use and wear and tear in the workshops the cartoons are in good condition. The strips of paper were pasted on a canvas, heads were restored. The weavers did not follow the cartoons exactly; more details were put in the foregrounds, the simple costumes were made richer, the gold design was added, some colors were changed; Christ's gown is white in the Miraculous Draught of Fishes on the cartoon and red in the tapestry. Raphael made several studies from live models

Homage to Goya

for several of the figures. The monumental and marvelous compositions were Raphael's own work in spite of other demanding commissions. The composition was sketched in charcoal which shows through the water-color-guash painting.

Engravings made of the series later were influencing several generations of artists and made this series to become the common heritage of Western Art

Las Hilanderas — The Spinners was painted in 1657 by Velasquez (1599-1660) and it is to be seen in the large Velasquez Hall of the Prado. It conveys the luminous silvery atmosphere of Velasquez's late period. It is a complex picture with several figures; it takes place in the tapestry workshop. The highly illuminated girl at right in cool greenish silvery tones on her white blouse, neck, face, arm is working, like the older woman at left in a shadow of penumbra. They are not spinning as the picture is titled mistakenly, but are preparing the wool yarn for the weavers, winding it on the bobbin. Accross the natural grey arch we see the next room; here royal visitors inspect a newly finished tapestry that is temporarily hung on the wall — it is a large tapestry, with figures, and a rich border.

Velasquez, court painter to the King, Philip the IV since the age of 24, painted Las Hilanderas only one year after his greatest picture, Las Meninas, the ladies of the court with the Infanta Margarita, the picture in which the painter also painted his self-portrait at the easel, with the aid of mirrors and three years before his great masterpiece, The Infanta Margarita, the delicate silver-rose changing values of light against his deep dark background. Velasquez, Marshall of the Palace since 1651, organizer of court festivites, decorator of various palaces and as such closely related to tapestries, died in Madrid exhausted after a trip to the French border where he had participated at the marriage ceremonies between the Infanta Margarita and the young Louis XIV of France.

Spain with its sense of grandeur, and the large rooms in the palaces was the ideal place for tapestries and till today there are great collections in the Royal Palace, 2,000 pieces, and the El Escorial; 46 tapestries of Goya, *Adventures of Telemaque,* Beauvais 17th Century, and such tour de force as tapestries after Hieronymys Bosch. During the 17th and 18th century Spain was more influenced by Flemish tapestry of the genre type than the French tapestries of the Kings' triumphs and pleasures, grandiose tapestries Paris produced at that time.

Philip V called in Jacob Van der Groten and his three sons and founded the Real Fabrica de Tapices in 1720. The Van der Groten family directed the royal workshops for 66 years.

David Teniers painted people, groups, animals, landscapes in informal compositions. Cartoons were copied from his paintings or changed and the design taken and adapted to tapestry. The Spanish looms took over the Flemish designs but soon there were Spanish subjects. Andrea Procaccini, Italian born Spanish court painter, made of series of *Don Quixote* with real racy native characters.

There followed a revival of French influence that swept over Europe. The court followed the French court etiquette and fashion, Louis XV chairs furnished the palaces and around 1774 French imitations of Watteau and Fragonard were models for the Royal looms.

The Spanish national character came back with the appearance of Francisco Bayen of Aragon, an artist of strong personality who found a happy balance of the influences of French grace, Flemish observation and Spanish subject matter. He had a good sense for decoration.

Goya. He was followed by his brother-in-law Francisco Jose de Goya y Lucientes (1746-1828), the great and brilliant artist who even in the field of tapestry, which was not his major work, far surpassed all before him and after in Spanish tapestry. Goya painted 45 cartoons with careless, easy haste — theatrical, exaggerated, picturesque — with a great feeling for decoration and the need of the wall. A prolific great designer, his tapestries bring joy, movement in the often stiff cold interiors. Goya is a transitional artist, marking the end of the Rococo and the beginning of a new area.

The tapestries are an early work; he worked for the Royal Tapestry Factory of Santa Barbara in Madrid from 1776; first tapestries designed for El Prado, later ones (1788-92) for El Escorial. Childish, charming games, scenes from Spanish country life in enchanting delicate coloring — the death of a worker by accident is the only instance where the laughter dies.

In 1786 Goya is appointed Painter to the King, in 1799 he becomes First Court Painter; he paints the magnificent large group portrait, Family of Charles VI and many official portraits.

Years later, tormented by deafness and later by political turmoil the artist paints the hunting murals in his house, Quinta del Sorde, outside Madrid. The pale rose and fresh green, the light luminous colors are gone; dark grey and black hallucinatory distorted images are the late period of Goya's oeuvre, this dark shadow doesn't disturb the cloudless blue sky of the Goya tapestries.

Folk Looms

Weaving Women Peru Singer

Africa

Peru. A small woven doll figure of a weaver from a grave 700 years old, is sitting on a pillow, in the University of Pennsylvania Museum's Peruvian collection. The whole genre scene, weaver, pillow, loom, and a tree, are 13 inches long and 18 inches high, made with great delicacy, have a robust charm, bright colors and beyond their artistic value contain valuable information for weavers. The pillow is loom woven fabric with a geometric design; it is stuffed with leaves and grass and sewn on 3 sides. The sitting figure of the weaving woman is also stuffed with leaves and made of woven fabric. She has black hair of wool, a shawl covers her head, her face is embroidered with black eyes, mouth and teeth, the square chin and nose in three dimensional modeling — a small textile sculpture, her arms, fingers are twigs covered with wool yarn, sticking out of a horizontally striped woven dress. Her loom is triangular, a frame of two twigs, tied on top and attached to the tree, widening toward the base and resting on the pillow. The frame is covered with decorative yarn of several strong colors. A few rows of weaving are started both at the bottom and the top of the loom; strings, bars are not sufficient for weaving. Branches of the tree are wound with cotton thread; there are blossoms, a bird's nest and a bird with wings and tail of woolen yarn of several colors and embroidery on the body.

A highly sophisticated fabric, a textile belt with bird design was woven on a very simple loom on the Central Coast of Peru about the 15th century. The American Museum of Natural History in New York has this loom with the unfinished weaving on it. Either the weaver or the person for whom the piece was being made died and the work left unfinished. Ancient Peruvian looms had only the barest mechanical essentials like two loom bars to which the warp threads were fixed, two sticks and a string to form the sheds. The museum piece is a backstrap loom but there were looms in the form of a frame or else pegs were driven into the ground and warps streched parallel to the ground — blankets, large pieces were made this way.

Peruvian fabrics were not cut or tailored. Had the belt been completed, it would have been 38 inches long with four finished selvages, but of this only 18¼ inches were woven; of wool and natural cotton in six colors. The geometric design at closer look reveals repeated stylized birds in diagonal bands with serrated edges. Each bird is woven of several colors in such a complex way that supposing there were additional sticks, cords — now missing — to produce the pattern, and a complex finger manipulation were used — even then we can't tell exactly how the pattern was formed. Peruvian woven fabrics from the uncompleted band of birds to large complicated compositions with many figures, animals, symbols in bright decorative colors and their peculiar angular style from the Archeological Museum in Mexico City to Washington, D.C. Textile Muesum, speak to us in a clear timeless language.

Peasant looms Far up in the rugged Carpathian mountains in a tiny isolated village of Transylvain folk art flourished untarnished by commercialism; women produced artistic and imaginative colorful woven cloth. The houses, pillars, doorposts, were richly carved with ancient magical symbols of the sun disk, bird-soul, tree of life, the pomegranate. The folk had a beautiful life; close to nature, full of poetry, songs, dance. Holidays were times when girls put on sparkling costumes. Everything was made at home; these proud people rejected

Weaving in Peru

and even despised machine-made products with which they had a chance to come in contact occasionally. They produced everything for their own use; like pottery, home spun cloth.

Keys and locks did not exist; instead they fabricated such complicated mechanisms, devices of hidden bolts out of wood, nails, cords, and only those who knew the combination could open the door. There were no two alike. On a warning when a foreigner approached they could lock houses, wells, fences, and exit on their backdoors through pastures and orchards, so that the unwanted visitor or dangerous element in this border region of a minority of Hungarians could find neither water nor food in the deserted village. If he was welcome, on the other hand, then he was feasted and doted with presents.

Spinning, weaving was a natural thing, a way of life; flax and hemp were produced and wool from sheep and goats were spun into yarn. Winter evenings, girls took their portable spinning wheel, treadle operated, and often a work of art by itself with carved, painted ornaments, into a barn. They cajoled an old man into telling them stories and kept spinning. Old women were spinning without the spinning wheel, even while walking long distances out to the fields or forest; with the staff under the left arm, parts of the raw wool pulled by the left hand feeding the right that turned the spindle. It is good to keep the fingers occupied in a creative way, it gives a balance of the mind; women did not smoke and did not need tranquilizers either. Girls worked for years on their trousseau, spinning, weaving, embroidering; it would have been a shame to buy these things. The finished pieces were saved in a tulip-decorated chest, and carried away in a procession on the girl's wedding day. Ornamental pillows were displayed at the pillow dance, danced by her unmarried girl friends.

Wool, cotton yarn of different thickness of several colors was used to make a heavy fabric that had stripes of colorful patterns; facing pigeons, birds, floral ornaments, geometric design were woven on a horizontal loom in a combination of linen weave and tapestry technique. Pillows, bedspreads, hangings, saddle bags and sacks were made of these fine samples of peasant tapestry weave. The sacks were beautiful with fringe and cords and were used instead of a shopping bag to carry things. The men constructed the looms; the uprights were richly carved wood pillars. Woolen skirts, linen blouses, some of which had small tapestry woven pieces sewn on, shoulder and neck pieces like coptic tunics. Lace, embroidery, every detail of their coloful costumes they produced themselves. The patterns were a heritage and I think they were born with this talent to make them because I often watched a peasant woman start to embroider a motif on a pillow in one corner without any outline or design, proceed and finish, months later, with the mirror image of the motif filling the entire space and derive from exact symmetry only that much as is agreeable and distinguishes hand products.

Magnificent were the beribboned woven and embroidered headdresses — peasant variations of a crown, usually made by the mother or grandmother of the girl; these were real works of art; tassels, pearls, tiny pieces of broken mirrors mixed media—were put togther with painstaking work; glass, artificial flowers appliqué — topped with dried flowers and ears of wheat. The folk did not feel deprived for the lack of certain materials; on the contrary, the limited

Navajo Woman

access to manufactured goods resulted in their great ingenuity.

Folk art died out not only as a result of war and social changes but when peasants were "discovered" at the turn of the century and their product put on the market — city dwellers' and customers' tastes soon had an effect — dealers provided them with machine made yarn and the time element dictated a more mechanical, repeated, or less detailed, less inspired work. As long as they made it for themselves or one beloved person it was art; once they had to make it—it was no longer the same.

Decorations on folk art objects, carvings on looms, woven pillows, have dates, initials. Folk art is not anonymous. The maker knows the object will survive him and in a way he assures his immortality. There is a continuity in the family as several generations use the same object; the objects are made to last. Often we find two names on a spinning wheel; "Joseph made it for Rosa;" a kerchief, "Mary made it for Andrew." Folk art is not only a collective expression of this visual heritage, there are talented individuals, whose work is original and outstanding. These are followed by lesser artists as usually occurs. Most articles produced had a practical use but products like the headdress, saddle bag, engagement kerchief with all their talent and virtuosity displayed were real *objet d'arts*.

Nomadic tapestry. In Bagdad caliphs lived in luxury, clad in sumptuous silks — while the poorest nomadic Arab families had goats and sheep and used the wool for home spun fabric. On primitive portable looms Nomads produced tapestries to decorate their tents, from the Sahara to Turkistan. There was no cartoon, this was the direct method; symmetrical repetition of geometric ornaments in brilliant colors and perfect craftsmanship. The design was handed over from generation to generations in memory — memory of the fingers rather than the intellect; with the secrets of dyeing the wool and the technique, which is somewhat different from the European one insofar as the threads are worked so that both sides are identical. There was a great variety in the details of the geometric ornaments and once in a while there was a personal little design; that of a dog, chicken, camel, horse, tent, that perhaps show how many animals were owned by the owner of the tapestry and also served as identification of ownership. The slits in the tapestry were left open; the tapestry was the wall of the tent, the slits let in the light through the intricate geometric patterns keeping the dimly lit tent cool and shady but permitting light enough for life to go on inside. For their delightful abstract design, bright color and great diversity and fine workmanship the nomad tapestries are most enjoyable.

At diverse points of the earth independently of each other tapestries are produced by primitive tribes, backward, isolated cultures. Under a cottonwood tree in Arizona a Navajo Indian woman is weaving a bright tapestry on a hand loom. A Yoruba woman in Nigeria sits in a palm thatched hut and makes home spun wool into colorful tapestry.

In India the ancient Charkha, a different type of spinning wheel is used in villages as in old times. For Mahatma Ghandi spinning was the symbol of national independence, the base of home industry in the over-populated country and a protest against English imported goods. Ghandi invented a folding spinning wheel, flat, with two discs for the use of men and women. His followers

143

Siberian Disstaf

made a vow of spinning every day; he used only hand spun and hand woven cotton fabrics. In his memory people still use the spinning wheel he invented but strangely it became popular among the educated men, the wealthier city dwellers while in the villages the old Charkha remained the prevailing one.

Handweaving in the United States of America. In the United Statees of America there is a great tradition of raising sheep, spinning and hand weaving. After shearing the sheep the fleece is sorted into grades of wool. Different breeds of sheep give different fleece and there is a difference between the fleece taken from different parts of the animal. The back has a thick, strong wool, greasy, it has to protect the animal in rain. The lower parts of the body are covered with softer hair. The fleece is then combed. "Carding" comes from combing with thistles; opening the wool and cleaning it of coarse parts and dust. After washing the wool is ready for spinning. The yarn will be of different neutral shades. If spun "in the grease" unwashed, the yarn is more waterproof and stronger. Mordanting and dyeing of the fleece or yarn can be done. The rough or smooth texture of natural fibers, their resilience have an influence on the woven fabric. How does a mass of shapeless fleece, unspun wool become yarn? Barbs unite the fiber and twisting multiplies strength.

From the arrival of the Mayflower there were several spinning wheels in use; in Colonial days the light "visiting wheel" was taken to a house where several women met for spinning. An interesting spinning method was the use of the great wheel or "walking wheel". These are the "how to do it" instructions: pull the wool, turn the great wheel, walk away from the spindle, change the angle of the yarn, thus made, walk forward still turning the wheel and return to it.

Hand woven fabrics were made among the Pennsylvania Dutch at the turn of the 19th century when flax culture was at its height. Small hand looms were made out of wood. Elizabeth Stauffer's weaving stool carries her name on it and the date, 1794, among painted and carved tulips and ornaments.

The Stauffers were a family of weavers of Lancaster County, later Chester County. An old account book records faithfully the weaving produced and payments received. Cost of a coverlet was $1.25 to $2.25 for the weaving; clients provided their own yarn. We quote from the account book of Peter Stauffer, born in 1791: "In the year of 1810, February the 11th, I weaved a ceiverled for Elisabeth Stauffer (charge) 8 (shillings)". At the age of 19 Peter Stauffer was already a well trained professional weaver. In 1812 he married Susan Hartz. By that time he was already settled in Chester County, Pennsylvania and busy with not only his loom but also other occupations like running a general store, general hauling and fishing for shad in the spring. Following years record the birth of 9 children and the purchase of an additional loom for $17.00. Probably several of the children were taught to weave, one of them, a daughter, in turn taught her son. We have three samples identified as woven by Peter Stauffer; one is a sampler piece in which the weaver tried out several variations of one star pattern; the other is a coverlet; the wool is red, light and dark blue with natural cotton yarn. The third piece, a coverlet is in the so called *Star and Diamond* pattern. Patterns were preserved in an old

145

German weaving book, the *"New Weaving Pictures and Pattern Book"* by Johann Michael Kirschbaum, published in 1771. There are patriotic patterns with George Washington, the word Liberty, the Eagle. Patterns are called charming poetical names like *Sorrell Blossom, Whig Rose, Snowflake, Cat Trail, Snail Track,* etc.

This past decade has witnessed a mushrooming interest in handweaving and spinning in the United States from Maine to California. Demonstrations, weavers at historical sites attract great crowds. Hand weaver craftsmen show their products at crafts fairs while art museums exhibit avant garde work by creative artists who have found a new means of expression in hand weaving and related arts. Craftsmen societies, guilds, workshops are growing, arts and crafts courses are taken by more students than at any time before. Spinning and weaving helps to relate to the past and is a very satisfying occupation. Hand woven products are greatly appreciated in our industrial society. Weaving is used to calm, to heal, sooth, to educate children, old people, the handicapped and has an increasing role in therapy with excellent results.

Textile Industry. While Elizabeth Stauffer still worked on her little loom Hargreaves had already invented the spinning-Jenny in 1770. Compton improved it and called it mule-Jenny, 1779. Dr. Cartwright invented the automatic loom in 1785. Horrocks invented the power loom in 1803 which is still in use.

Jean-Merie Jacquard (1752-1834) silk weaver of Lyon invented the machine named after him. The work consists of mechanically selecting and lifting the warp threads when the shuttle passes accross the loom ; the action is regulated by means of cards with holes through which the lifting needles pass. The holes are an early forerunner of today's computers.

But even before the mechanization of spinning-weaving there was a thriving textile industry during the 17th century in England and the United States. By that time guilds could no longer control the field; high entrance fees and town regulations kept many a craftsman away. Unprotected country labor was cheap and widely used. Workshops of textile industry were later established near rivers for water power; with Watt's steam engine large amounts of water were needed for steam boilers and the river was used to dump waste material into it. What happened to the idyllic occupation of spinning-weaving — it became the source of all evils. Man's outlook to work has changed. The workers in the guild, the apprentices under the master weaver had a personal pride; their work was their life and passion and inner necessity, even if over-production reduced prices they were working with pleasure. Now the relationship to work became impersonal; inhuman low wages, windowless depressing big buildings, monotonous work and slums were signs of this change.

The more the production of textiles has been mechanized, the more it became a mass product, the higher the prestige and value of hand woven fabrics and the king of all weaving techniques; tapestry in which art and quality are highest.

Le Chevalier tapestry by Domjan. Right from the Hungarian plains this proud horseman rides into the world on his blue stallion. He himself must be an outlaw — one of those old time romantic legendary figures with a gun in one hand and a peacock feather in the other — violence and poetry. The horse, beribboned, of strange proportions, large head, long neck, red-eyed, dishevelled, carries his rider proudly. The tapestry is strong and decorative, a play of blue, red and purple that gets a unity from the golden sunshiny brilliance that flows over all shapes against a dark background of the night.

Morning. Two large superimposed leaf forms; the full skirt and; the big tent of a richly fringed shawl above cover the woman figure; her small pale face and hands disappear in the strong ornamentation of the dress, rose motifs in positive and negative and in contour lines, a strong linear system of lines in dark blue-black make out the shawl; abstract ornament of embroidery fill the skirt. The tapestry has a plain background, only the fine battage of close blue-purple shades offer a variety. There is a small bird in the upper right corner — bird of the soul always present. The tapestry is of muted silver-grey-blue colors with darker color for the design.

Holiday Wine. A lovely girl offers a bottle of wine; the charming small head (pale blue) is ornate with dry ears of wheat, and fluttering ribbons in shades of pale rose embroidered in old gold. The fringed shawl of light turquoise has a linear design in green-gold. The bright red skirt is decorated with bright yellow ornaments in a linear design of plant motifs and butterflies that grow on a stem and continue growing out of the skirt in pale old gold on the dark green background. The pale delicate shades of rose-ivory of her many petticoats are ruffled echoing the ribbons above. The fat little legs end in tiny feet in laced high shoes; the idyllic picture of the peasant girl in her holiday costume is of times gone.

For a subject matter *Holiday Wine* is the symbol of all good things in life. As a tapestry, it is a beautiful balance of brilliant red and light turquoise colors, clear, decorative. The red wine in her wine bottle in the upper left corner of the tapestry is the center of the color composition.

Dance. Two figures from a fairytale world swirl around; their beauty is intoxicating. The girl in love is glowing from an inner radiance, a peach to coral to carmine hue covers her figure, the radiance reflects on the angular shaped boy figure in shades of gold. Nature is present in floral and animal motifs; deer, rabbits on the boy's cloak, birds in the girl's ribbons, headdress — a large peacock in her swirling skirt, leaves, branches, thistle on the boy. All these side motifs barely suggested in small dots and lines, peacock feathers, and maiden hair fern decorate the boy's hat and fall back in a wide curl, ribbon, kerchief flutter around. There is movement in the broken-up lines of the green background, battage of lighter and darker green. The shapes turn, move, melt, the colors sing and dance; every inch of this tapestry is a gem.

Spinning Wheel

Tapestry in the Home

Liberty

Mother and Child. This mystic, eternal mother or fertility goddess in pale yellow light—colors of golden harvest, ripe corn, luscious fruits—is surrounded by an ethereal aura of blue. The yellow is actually her bell-shaped skirt, tinted with many hues of earth colors; on it in darker purple design there are flowers—tulips, carnations, pomegranates; a large and elaborately embroidered bird—the phoenix—symbol of rebirth, and the rhythmic renewal of seasons; sun-symbol makes us think of seeds buried in the earth, in the dark, that germinate, reach toward the sun, seeds germinating, children conceived.

The child is hardly visible, still one with the mother-body. The flowers continue blooming outside the shape of the figure, and in the corner there is a departing shadowy bird in blue on the darker blue background, right under her skirt. A large fringed shawl of soft pale blue-grey like a dome covers mother and child. The facing large mother figure is symmetrical, motionless, like an idol—a timeless image.

When Domjan makes a sketch, a shape, an outline, it is for the beauty of the line. The motifs and the details get into it during the process of cutting the woodblock; they are not preplanned, but happen by chance. Many months later the colors are created—in the heat of creation—again, not thinking of narrative contents, iconography — and so a work in born. Then later a title is found, which at best, may relate to the visual creation. Whatever was said in case of Mother and Child—or any other tapestry—symbolic meanings, reasons, were certainly not planned that way and may or may not have been in the artist's mind subconsciously while creating the work. When a color woodcut is created, finished, it is a stranger. When a tapestry arrives, with all the work that went into preparing it for weaving, it, too is a stranger. Nothing can be changed. As long as he is working on it, the artist is one with his work; once the work is finished, the umbilical cord cut, the work takes on a life of its own. Here it is and that is how it is. Much later, it reveals itself, unfolds, and offers its treasure. Certainly, there can be many interpretations. Close as I am to this creative process, for me as for others, the art continues to be a mystery.

* * *

The acquisition of a tapestry is an important moment of destiny. Whether it was the result of love at first sight, or else of careful considerations of it as an investment or as part of the home interior, there evidently was a great desire, a passionate emotion, even physical attraction to the material itself — and a feeling of triumph as the deal was closed, and the proud possessor at last could enjoy the tapestry with his family or in solitude. The owner tries to get more information on tapestries in general and the history of his own tapestry in particular, like a lover who wants to know all about the beloved; family and friends share his curiosity; a tapestry can talk to a broader group of people than paintings. ·

The tapestry looked attractive in the gallery or exhibition, but only in his own home has the owner the chance to contemplate the tonalities, the finesse of design. Only then can he analyze the hidden meanings and symbolism, and penetrate into the designer-artist's thoughts, psychoanalyze, if you like, and find the key to the work.

A tapestry can reveal its beauty at any hour from morning to night, in natural or artificial light, as well as in the changing light conditions through the seasons. The more the owner looks at it the more enchanting details he can discover — till he knows the tapestry by heart. By this time the tapestry, first a stranger, has become part of the family; he could not live without it. The tapestry has become a friend that can never hurt but will always console and share in his joy.

To look at a work of art in intimate contemplation, to enjoy it is a creative process, one has to exercise to develop a cultivated mind and sensitive eye that can absorb its beauty. There is a tremendous difference between walking through a museum and saying it was nice, and looking at a certain work and suddenly feeling that you have been struck by lightning. Beauty burning through, feet had taken root. The intensity of feeling can perhaps be measured on an electronic meter; a certain art work burned into the memory, unforgettable.

Exercise in meditation. To look at a tapestry, sit motionless at a comfortable distance, eyes directed at the center of the composition which is not identical with the geometric center; with eyes at rest within the limits of the tapestry, going around then we select one particular small shape and here the eyes are stopped and kept voluntarily while we stop the half-conscious constant flow of thought—chain of ideas—memories, worries, fears, tension, that pass through our mind all the time and we direct our thoughts to the tapestry; other thoughts, pictures are shut off. I have selected *Frosty Night* by Domjan, a square-shaped tapestry of rather muted tones, for this exercise. The first glance registers the red eye of the deer, the center of the composition; the graceful neck; long head, turned around; two large antlers of straight, pointed shapes; small body; front legs crossed; the contrast between curved and straight lines, between the warm of the red eye and the cool blues of the rest; pale silver-grey on the dark background.

With eyes still focussed on the tapestry and the rest of the world forgotten, we stroke this gentle soft silky deer with our glance, starting with the antlers. We discover more small running deer, rabbits, animals in the grass. We notice the deer has a small plant particle in its mouth, we follow the neckline to the body and find more small animal shapes, squares, triangles, we follow the graceful movement of the legs and I find myself sitting cross-legged. We see finely cut leaves and a stem under the legs; we follow the stem upward to the flower—a Queen Anne's Lace, I think—then another on top and through the scattered faded dots and lines around return to the red circle. Here the eye stays. The title suggests winter and North, but then how come there are flowers? One is puzzled; it is a paradox; is the deer resting on a rock daring a Nordic summer, or is it seen across ice formations in shape of a flower — is it a winter evening, blue misty shadows of the forest nearby; or else the shadowless light of the midnight sun, pale distant skies, silver moss?

Another strand of thought runs in another direction; one may ask, is this a tapestry; how was it made? In one's mind follow the thread in its journey up and down, and reversed on the way back; the battage, the insertion of a thread of another shade—how many fine nuances, values? With what fast and precise movements exercised fingers moved rhythmically? How many times was the

151

color changed then tied in the back? How many times were the rows tightly pushed together with the comb? How did the tapestry grow, row after row? There were tight parallel cotton threads first naked—stretched out on the robust wooden loom; till they got all covered with the colored wool threads. Also the cartoon, painted on paper behind; the wool was dyed, those gentle close shades — the red from where did it come? Large chemical factories, from nature, the earth; then the wool, before it was wound unto the bobbin, the yarn was spun out of colorless soft bunches of virgin wool; the sheep was sheared with scissors, bouncing on the pasture, light, when released from the strong grip of a man, the wool falling off, gathered in bags, still greasy, dusty, and the shepherd . . . How much human labor, care, skill and knowledge was necessary to produce the tapestry? Men we never saw, from the shepherd to the weavers, did their task with care, worry and joy. Faraway places were involved; the wool in the heat at the bottom of a cargo ship; the unloading, the train or truck and the toy train of Aubusson that comes twice a week; the airplane that brings the finished tapestry; customs, insurance, official things, papers to fill, serious financial matters involved around a creature from the land of dreams. The tapestry on exhibit, all the people who saw it, and kept its memory . . . and now it is here in the silence of the home and telling us stories.

Thoughts continue—before it was a cartoon, there was a color woodcut; a famous one that had been loved, admired, and reproduced in books, and exhibited. How was it made? Printed on a special paper with hand-made colors of pigment powders, all those pigments, from where did they come? The earth colors called from their places of origin; Sienna earth, Pompeii red, organic and chemical pigments, intense pure colors crushed on a marble slab, then mixed with oils and French turpentine — and built upon the block, the brush puts them on the block — then with the pressure of the finger applied on the reverse side of the paper the color is transferred from the block to the paper—there are many superimposed layers of colors—is it going to rain again? The colors don't dry properly — it has to wait for the next layer of colors—or else it is too dry and the work can't stop all night. But before the printing there were wood-blocks, first one, cut with sharp knife and gauges on a plank-wise block the design emerged; a finished shellacked wood-block, a piece of sculpture, bas-relief. The wood-block, the color woodcut print, the tapestry; the transposition from one art work into the other; this complex, intricate set of methods, detailed, delicate, each a work of love; and dedication.

The wood-block, from where did it come? It was a tree growing from the earth, a seed first, then two tiny leaves, a root reaching down, a stem stretching upward to the sun, then many leaves, rains, wind and snow, and sun; many years, till the trunk grew, the tree aged, was cut down, the trunk resting eight years—

Think of the sounds; the shepherd with his flute, the dogs; the woodcutter, then the art work, cutting of the background with a little rubber hammer; the weaving workshop, rhythmic sound of the pedals, even like weaves, the chatter of the young girls—and think of the time element . . .

And the artist — before the first tentative light line was made, what did he think? What made him do the deer? Where did it come from, this idea, of the silver-blue deer among flowers, which line was the first on the virgin paper, or was it a wood-block? . . . And so it continues and another time the exercise would be to select a dot of the design, hold it with the eyes and empty the mind of all thoughts in peace so that a ray of divine harmony can fill the void . . .

Shephred   Sonzen

Evelyn Alexandra Domjan, a one-time student of Joseph Domjan's at the Academy of Fine Arts in Budapest, his wife, mother of their three children, and a life-long collaborator in the woodcut work first won recognition for her art work by nation wide competitions at age six. "As soon as I could walk, I was taken to the museums" she remembers.—Although she was educated in latin school and at home in music, literature, foreign languages, and the arts, according to traditions of European intellectual middle class, her preference for the arts was immediate. She learned early the satisfaction derived from creating art and developed the taste and pleasure of looking at art created by others to the point of extasy. It is a fact that she fell in love with Domjan's early color compositions years before she met the man. Sharing the dramatic life of the artist with its lights and shadows—to be rich or poor—made no difference to her since her basic and urgent need to be surrounded by beauty was always satisfied.— Medieval arts, Oriental arts have taught her to be proud of being a nameless member of the universal brotherhood of artists-artisans that reaches through the centuries.

She plans to continue her research work in museums, her work in the woodcuts, her garden, and her writing.

She shares Domjan's motto: "Art is my life".

Spinning   Domjan

Love Birds

Washboard   Donijan